We, robots

Lode Lauwaert • Bartek Chomanski

We, robots

Questioning the Neutrality of Technology, Ethical AI and Technological Determinism

Lode Lauwaert
Institute of Philosophy
University of Leuven
Leuven, Belgium

Bartek Chomanski
Department of Philosophy
Adam Mickiewicz University
Poznań, Poland

ISBN 978-3-031-77173-6   ISBN 978-3-031-77174-3 (eBook)
https://doi.org/10.1007/978-3-031-77174-3

Translation from the Dutch language edition: "Wij, robots" by Lode Lauwaert © Author 2021. Published by Lannoo Campus. All Rights Reserved.

© The Editor(s) (if applicable) and The Author(s), under exclusive license to Springer Nature Switzerland AG 2025

This work is subject to copyright. All rights are solely and exclusively licensed by the Publisher, whether the whole or part of the material is concerned, specifically the rights of reprinting, reuse of illustrations, recitation, broadcasting, reproduction on microfilms or in any other physical way, and transmission or information storage and retrieval, electronic adaptation, computer software, or by similar or dissimilar methodology now known or hereafter developed.
The use of general descriptive names, registered names, trademarks, service marks, etc. in this publication does not imply, even in the absence of a specific statement, that such names are exempt from the relevant protective laws and regulations and therefore free for general use.
The publisher, the authors and the editors are safe to assume that the advice and information in this book are believed to be true and accurate at the date of publication. Neither the publisher nor the authors or the editors give a warranty, expressed or implied, with respect to the material contained herein or for any errors or omissions that may have been made. The publisher remains neutral with regard to jurisdictional claims in published maps and institutional affiliations.

This Springer imprint is published by the registered company Springer Nature Switzerland AG
The registered company address is: Gewerbestrasse 11, 6330 Cham, Switzerland

If disposing of this product, please recycle the paper.

Contents

Introduction ix
1   **The Neutrality of Technology**   1
   Having and Being   3
      Decolonizing Technology   4
      The Grandfather's Watch   5
      Moral and Not Moral   6
   Technology and Ethics   8
      Crying About a Robot   8
      Without Moral Value   9
      Capitalism and Religion   11
   Technology Is Neutral   13
      Watching Temptation Island   13
      Everything of Value Is Visible   14
   The Value Ladenness of Technology   15
      Survival and Living Well   16
      The Moral Duty of the Designer   16
      British Telecom's Phone   17
      Improvement Through Technology   18
      Moses' Mistake   19
      Repulsive Things   20
   Some Critical Notes   21
      The Neutrality of a Drill   22
      And Justice for All   23
   The Importance of Neutrality   25
      Seeing Values   25
      Designing and Using Technology   27
      It's the Economy, Stupid!   28
      Machines Are Not Rocks   30
   The Point of a Philosophical View   32
      Praising Human Beings   32

	The Moral Relevance of Knowledge	33
	Taking Care of Things	34
	Moral Machines	34
	Conclusion	36
2	**Ethics of AI**	**39**
	The Algorithmic Society	40
	Oral-B's Toothbrush	41
	Move Fast and Break Things	42
	Disruptive Technology	44
	Moral Disruption	44
	Law and Ethics	46
	An Ethical Look at AI	47
	Rage Against the Machine	48
	Bad Use	50
	The Panoptic View	51
	Racist and Sexist	52
	Safe and Secure	55
	Like a Dark God	57
	It's the Ecology, Stupid!	59
	The Seventh Sin	62
	Autonomous Cars and Weapons	62
	Nobody Is Responsible	64
	What Is Responsibility?	65
	Tour of Duty	67
	All Hands on Deck!	68
	Can Robots Suffer?	70
	Child Soldiers and Drones	73
	Conditions of Responsibility	74
	Everything Under Control	75
	Breaking Pot, Paying Pot	77
	Acting Under Coercion	78
	Conclusion	79
3	**Technological Determinism**	**81**
	From Printing Press to Protestantism	82
	The Technological Condition	84
	Determinism Determined	85
	Colonizing the World	87
	Heidegger and Fascism	87
	The Essence of Things	88
	The Technological Imperative	89
	Technologization and Technocratization	90
	From Heidegger to Google	91
	Beyond Utility	93
	The Problem with Instrumentalization	95

Types of Technological Determinism	98
Emergence and Development	98
Don't Shoot the Messenger	99
In the Grip of Experts	101
The Social Effects of Technology	102
Are Social Media Polarizing?	102
Call Me, Write Me	104
Necessary? Possibly!	105
The Birth of Technology	106
Twice the Wheel	107
Simultaneous Evolution	108
Coincidence and Necessity	109
The Evolution of Technology	110
The QWERTY Keyboard	111
The Steam Engine	112
In the Line of History	114
Patterns in Development	115
Moore's Law	116
Inevitable and Responsible	118
Technology as a Social Construct	119
Starbucks' Schedules	120
TINA, There Is No Alternative	121
The SCOT Approach	122
Between Dream and Deed	124
Constructivism and Activism	127
Conclusion	130
Afterword	131
Bibliography	133

Introduction

Machines are taking our jobs. We all heard this. But sometimes, machines will also be the ones who tell you your services are no longer required. Stephen Normandin, an army veteran, is a case in point. During his service in Arkansas, he prepared meals for Vietnamese refugees resettled in the USA. For the last 4 years, he had been navigating the streets of Phoenix, delivering packages for Amazon. But his tenure came to an abrupt end recently. The technology employed by the multi-billion-dollar corporation, which kept a constant watch over him, determined that he was not performing his duties adequately. The verdict was delivered not by a colleague or a supervisor, but by an impersonal, automated message: Normandin was asked to leave. This stark decision, devoid of human judgment, was the cold conclusion drawn by the unforgiving algorithms of modern technology (Soper, 2021).

It is not just the jobs. Both of us, when we wake up, tend to look at our phones first. You may think: "So what? Don't we all?" This statement is stunning in its commonality. An estimated half of those who own a smartphone start scrolling in bed in the morning. The morning routine of millions of people has drastically changed over the course of not much longer than a decade—because of a small rectangular box. About that—phones often measure 13 inches in length, hardly unusual in an era where the typical smartphone stretches nearly 14 inches. This size is particularly suited to male users, aligning closely with the average length of a man's hand, which is generally longer than that of a woman. This design subtly embodies a masculine perspective, ingrained in the very dimensions of our modern devices (Perez, 2019).

On the last day of the year 2019, a red light went on in the rooms of Canadian software company BlueDot. Based on the analysis of countless messages, the company's technology concluded that an outbreak of the coronavirus SARS-CoV-2 would occur in the Chinese city of Wuhan. When it became clear that we were in a pandemic, Chinese companies rolled out facial recognition technology to detect citizens who were not wearing face masks. In numerous countries, apps were developed for citizens to download for free that signaled whether they had been in close contact with an infected person. In addition, technology was deployed to detect and combat the virus. An algorithm was developed that could indicate with a high degree of probability in less than a minute whether someone was infected or not. Google then used its technology to find out what components the virus is made up of—successfully, by the way. It discovered that the so-called 'spike protein' was the secret weapon by which the virus binds to body cells.

From corona to colonialism. Bangalore, often heralded as the 'Silicon Valley of the East', epitomizes India's vibrant focus on communications and information technology. Yet, India's technological prowess has deep roots, significantly influenced by nearly two centuries of British colonial rule beginning in the mid-eighteenth century. In 1819, for example, the Indian populace witnessed the launch of the first steamboat on an inland river, a luxury more symbolic than utilitarian, gifted by the British to a local prince. This era also saw the British orchestrate the development of a sprawling railroad network across India. By the time India gained independence from the British Empire in 1947, the country boasted approximately 70,000 km of railway tracks. Another significant technological advancement under British rule was the telegraph. Initiated in the mid-nineteenth century, the first line stretched over 1000 km between Calcutta and Agra, quickly followed by submarine cables linking England and India. Clearly, these developments were strategic, designed not to benefit the Indian people but to enhance British control. The railways expedited troop movements, the telegraph improved communications with Britain, and the maritime routes facilitated the export of vital raw materials such as iron and coal to England (Basalla, 1988).

Today, bacterial infectious diseases do not significantly impact mortality rates as they once did. In earlier times, before Alexander Fleming's discovery of penicillin in 1928, these diseases were the primary cause of death. However, there is a looming risk that they might become a major threat again. This is due to an increase in antibiotic resistance, exacerbated by the excessive consumption of these drugs over recent decades. Against this backdrop, recent research from the Massachusetts Institute of Technology shows promise. There, researchers trained an AI system with the structures of thousands of antibiotics to identify the chemical structures of effective molecules. The AI then evaluated 6000 substances and singled out one particularly promising molecule: halicin, named after the computer HAL from Stanley Kubrick's 1968 film, *2001: A Space Odyssey*. The computer predicted that this molecule would be effective in fighting bacteria. Intriguingly, halicin's chemical structure is unlike any existing antibiotics, marking a significant breakthrough in the field (Trafton, 2020).

Finally, there is a strong connection between technology and ecology. Some technologies—smartphones, for example—last only a few years, but at the same time are profoundly marked by ancient natural history. In the mid-nineteenth century, for example, a tropical tree enjoyed the interest of the tech world: the *Palaquium gutta* from Asia. The reason was that the sap from this tree contained a substance called 'gutta-percha.' This substance was useful because it insulated the transatlantic telegraph cable at the bottom of the ocean, the cable that allowed communication between North America and Europe through Morse code. The problem was that the first cable required 250 tons of gutta-percha, each ton of which required 900,000 strains of *Palaquium gutta*. In Asia, several jungles disappeared for this reason alone. Similar issues continue to be at play today. The production of iPhones, drones, motor vehicles, batteries, and camera lenses requires numerous minerals that are obtained from Indonesia, Congo, and Mongolia, among other places. The supply is scarce and if the minerals were no longer available, the growth of technology would stagnate. It should therefore come as no surprise that a Tesla Gigafactory was built not far from Silver Peak in Nevada, where lithium is mined. Lithium is the raw material used in the production of batteries; the batteries of Tesla's electric cars require large quantities of it (Crawford, 2021).

Technology and AI

This book is about technology.[1] We talk about different types of technology and pay particular attention to artificial intelligence (AI). Thus, the book will be about smart and dumb things: trains, steamboats, apps, airplanes, robots, watches, bicycles, drills, planning software, bridges, coffee machines, vibrators, cotton harvester machines, nuclear power plants, phones, social media, electric barbed wire, toothbrushes, cars, and computers.

There are a number of good reasons to pay attention to them. First, technology is an integral part of our lives, to the extent that there are very few, if any, things that we do without using artifacts made by designers and computer scientists. Second, quite a lot of people do not know exactly what AI is and whether this technology is present in their lives. Third, technology is responsible on the one hand for many problems and misery, and on the other for the increase in prosperity and well-being. Industrialization created black smoke and black rivers, but medical technology has allowed us to prevent, detect, and cure diseases more quickly. Fourth, governments and companies invest a lot of (public) money into the research and development of technology every year. Fifth, the tech industry is one of the leading and most capitalized on sectors of our society. In addition to Big Finance and Big Pharma, there is also Big Tech, especially the Big Five: Amazon, Google, Microsoft, Apple, and Facebook. And finally, technology is as old as the human species itself. The development of genetic engineering and the self-driving car were preceded by the making of fish hooks and nets, and by the fashioning of knives from horns and tusks.

Some may find that last comment remarkable or even unjustified. Are fish hooks and nets technologies? To what, in other words, do people refer with the term 'technology'? What do we actually mean when we say that something is technology or, more specifically, AI? It is important to reflect on this, because in order to avoid misunderstandings, it must be made clear what we are and are not talking about.

---

[1] This book is the translated and slightly reworked version of Lauwaert (2021).

The Meaning of Technology

To begin with, we would like to point out that the word 'technology' is linguistically related to numerous words that nevertheless have very different meanings. We can say that the Dutch footballer Jaap Stam was not as *technically* gifted as Diego Armando Maradona, who entertained the audience with a few *technical* feats during the warm-up for the match between Napoli and Bayern Munich in 1989. The legendary Argentine did this to the music of the band Opus, music that cannot be compared to the *techno music* of Ken Ishii and that would sound completely different without the help of a sound engineer. Other familiar words leaning against 'technology' are 'technocracy,' 'technicity,' 'technoscience,' 'technologization,' 'technophilia,' and 'technophobia.' The common thread running through this long enumeration is 'techn-,' which goes back to *tekhn* from Sanskrit. That root, more than 4000 years ago, referred to things such as woodworking and carpentry.

In addition, a glance at history reveals that the word 'technology' has been interpreted in various ways over the past 2000 years. This was not the result of a shrewd linguist or philosopher but simply characterizes the use of the term 'technology' by millions of people throughout the ages, just as is the case for 'science,' by the way, and also for 'fairness' and 'responsibility,' for example. We deal with the latter later, and for now focus on the multiple meanings of 'technology.' At least two of them are hardly used nowadays. Which ones?

The oldest meaning of 'technology' dates back to Classical Antiquity and had all but disappeared by the end of the nineteenth century. It means that technology, like psychology or sociology, is a science. More than 2000 years ago, technology was a study of liberal arts such as grammar and rhetoric. Later, from about the eighteenth century, the study of illiberal arts or manual labor was also called 'technology.' It then also meant the study of, for example, cooking or working with machines.

The second meaning is less ancient. It can be found, for example, in Karl Marx's *Das Kapital* from 1867. In several places in this book, he uses 'technology' in the sense cited above, namely as *Wissenschaft*. But he also thinks that a history of technology should be written, and asks the rhetorical question whether the history of the emergence of man's productive organs in society, of the material basis of each individual social organization, deserves equal attention (Marx, 1970). This shows that for the German philosopher, as well as for several of his contemporaries, 'technology' has a second interpretation. Namely, it also refers to the production process.

Nowadays, when we say 'technology,' we do not refer to something like science. Usually, we have two other things in mind: a process of manipulation, and a material thing. First, by 'technology,' we mean the manipulation process involved in biotechnology. The term 'biotechnology' itself has not been used for a particularly long time, but it does refer to a process that is centuries old. Think of animal and plant breeding, the process of selecting and crossing desired traits with the goal of keeping existing animal and plant species in an improved condition. Genetic engineering also refers to a manipulation process. In this case it is a process of transferring a gene that codes for a certain characteristic from one organism to another, so that the genetically modified organism also possesses the desired characteristic. Second, by 'technology,' we also mean objects such as those listed by Ewan McGregor at the beginning of the song 'Choose Life' from the 1996 British cult film *Trainspotting*: televisions, washing machines, cars, CD players, and electric can openers. To this list we can also add toasters, irons, electric toothbrushes, computers, cameras, and medical devices such as infusion pumps or pacemakers.

When we talk about technology from now on, we mean these two things: either a manipulation process or a thing like a smartphone or car. Although we should immediately add that it will usually be about the latter. We *never* talk about technology in the sense of science or the production process, *sometimes* we talk about biotechnology or genetic engineering, and *mostly* we talk about things.

We assume that this is consistent with how most people typically use the term 'technology.' Nevertheless, we would like to draw attention to the following. Everyone or just about everyone may generally refer to things as 'technology,' yet there is no unanimity on exactly which things are and are not technologies. Certainly, particle accelerators and smartphones may be technologies to everyone, just as computers and defibrillators are. But for some, a drill is a technology whereas for others it is not; one person considers a radio to be technology whereas another does not. These differences are related to other differences, for example, differences in origin or age. For example, if you are an engineer, then chances are that you do not consider a drill to be a technology, but if you are an anthropologist, then you may consider furniture and garments to be technologies.

So it is advisable to be clear about what we do and do not consider technology. We take a broad view. Broad, in the sense that we include under 'technology' a whole range of things: not only computers, cars, and AI but also drills and furniture. We choose such a broad interpretation because there is actually no good reason to call a computer 'technology' and not a chair. Of course, there are differences. One technology is based on scientific knowledge, it is made of silicon and has buttons, whereas a chair normally does not have any of that. However, differences such as this are not relevant to distinguish the two things so strongly from each other, and to apply the story that will unfold in the coming chapters only to, say, computers and not to, say, drills.

A Broad View

The past few comments are sufficient to give a rough definition of 'technology' as many people tend to use the term, and as we do. That definition is that all technology is artificial in nature, equipped with a function and material in nature, or closely related to something material. Let us clarify that.

When we say that something is artificial, we usually mean that it is a result of human action—usually, because 'artifact' is sometimes also used when talking about animals that are not humans; but that is not important here. This description is still too general, however, because enlarging the hole in the ozone layer is also an effect of human intervention. Humans are thus causally connected to the creation of an artifact, but that is not enough to see something as an artifact. Hence, this refinement: an artifact is the *intended* effect of human action. Suppose, however, that one plants a seed in a garden, with the intention of growing a tree. When that actually happens, that tree is the desired consequence of the action of putting the seed in the ground. Yet we would not say that the tree is an artifact. That is because an object of nature can grow even without the intervention. So, things of an artifactual nature are things that come into being (almost) solely on the basis of human intervention. Note that artifacts differ from objects of nature, but may be composed of those objects. A canoe is an artifact even if it is made entirely from a tree trunk.

So, all technology is artificial, but not every artifact is a technology. There are other things that are artificial, but about which few, if any, would say they are technology. Think about music. It is not exclusively sound, but the result of at least one person's plan to arrange sound in a certain way. So, music does not come into being the way in which a tree comes into being, yet we do not usually think of it as technology. This is because technology always has a function in addition to being artificial. That is, all technology is designed with a purpose in mind; without that purposefulness, a design is not technology. Music can obviously be useful, but music that has no purpose is still music. Utility, in other words, is not crucial for music, whereas it is for technology. Of course, we also know that technologies sometimes temporarily or permanently stop working, but that does not mean that they are no longer technologies. Something that is human made is a technology if it is designed to accomplish a goal. This remains true even if it turns out that the design no longer does what it is supposed to do, temporarily or otherwise.

Finally, virtually all technologies are material, in the sense that they are measurable and tangible. But this is not necessarily so. Chatbots, for example, are technologies, but one can neither feel nor weigh them—although their output is measurable. On the other hand, there are no chatbots without the existence of something that has material density: hardware. Hence, this description: although technology is not by definition material, it is always connected to something material—whether that material is steel or iron or otherwise is irrelevant.

Technology differs in this respect from legal laws, such as the law against discrimination. Such a law is clearly an artifact and, moreover, made with a purpose in mind: to protect people or groups. Yet a law is not a technology: the former is not attached to a physical entity, whereas the latter is. So are bills and coins technologies? They are made by people, designed with a purpose in mind, and have a material character. That is true, but there is no close connection between the material nature of a coin and the function of a coin. That a coin is money is solely a matter of convention. A lot of objects could in principle be money—coffee beans, candies, or dice. In the case of technology, however, there is a strong connection. A technology can only realize its intended purpose on the condition that it has this particular shape, has that particular weight, and so on.

General and Narrow AI

We find ourselves asking another big question: what do people really mean when they say 'AI'? There are three main ways of thinking about it. First, many people use 'AI' the way in which we use 'intelligence' when talking about humans and animals. In this sense, intelligence is something like a skill that humans, or even dogs, might have. From this angle, AI is just like that skill, but for a computer or a robot instead of something alive. Second, there are those who use 'AI' to mean a whole area of study where scientists research various ways of creating smart machines. This kind of work happens in universities and private laboratories. However, most of the time, when people say 'AI,' they are not talking about a science or a special ability. They are usually referring to a type of technology that can do things that other tech cannot. That is the version of 'AI' that we are going to stick with here.

Obviously, not all technology is AI, but in our view, given the above, all AI is technology. That means that some of the things that we discussed a few lines ago also apply now. All technologies are materially embedded and so is AI. Smart systems need hardware; there must be computers to process large amounts of information and to do fast calculations. In addition, there is no single answer to the question of exactly what technology one has in mind when talking about AI, just as there are multiple meanings of the term 'technology.' That has to do, among other things, with the distinction among these three things: superintelligence, general AI, and narrow AI. Let us explain that distinction.

The expression 'artificial intelligence' was coined by computer scientist John McCarthy in the summer of 1956 during the famous Summer Research Project on Artificial Intelligence at Dartmouth College in New Hampshire, USA. Today, the phrase is very occasionally used to refer to artificial entities that far exceed the capabilities of humans: superintelligence. It captures the imagination, and although it is this type of AI that regularly appears in movies and popular media, it does not currently exist. It is questionable whether such technology will ever exist, let alone whether we know when it will be created (Bartoletti, 2020).

The same more or less also applies to what is often called 'general AI.' This is technology that is closely related to how we humans function. It is called 'general' because it is able to perform more than one task, simultaneously or not, and to use the information from one task for another (Lambert, 2019). But as indicated, such technology does not currently exist, even though there are currently numerous companies, designers, and computer scientists around the world working to create such a thing. General AI is pretty much the dream of many technologists (Tegmark, 2019).

Today, when politicians, researchers, and others praise or criticize AI, they almost always mean the third variant: narrow AI. It is also the only form of AI that currently exists and is being used. One speaks of 'narrow' here in the sense that the technology is only capable of performing a certain specific task, and cannot switch between different types of tasks. This is a limitation, but at the same time gives it the advantage of being very good at what it is made for—and not seldom better than humans.

Narrow AI has been used for a variety of tasks: recognizing the faces of people and other animals, determining the likelihood that a prisoner will reoffend, processing natural language, plotting the shortest route to your destination, estimating where violent crimes will be committed, predicting the results of sports competitions, spreading fake news, showing commercial and political advertisements to the right target audience, estimating whether someone is creditworthy, locating terrorists, determining the risk of a virus outbreak, detecting cancer, recommending music and television series, setting insurance premiums, suggesting potential friends and partners, selecting job candidates, tracking and evaluating employees, and much, much more (Russell, 2019).

One can further organize everything related to narrow AI by underlining the distinction between two types: expert systems and machine learning. What exactly does that distinction entail?

The first type of (narrow) AI, often referred to as Good Old Fashioned Artificial Intelligence, popped up in the second half of the twentieth century. Its goal is pretty straightforward: to make decisions, evaluations, or recommendations based on the data that it receives. Consider MYCIN, one of the pioneer expert systems created at Stanford University in the 1970s. It analyzed patient data to assess whether the patient might be suffering from a blood-clotting disorder or a bacterial infection. The crucial part happens between the input (the data) and the output (the decision): the data must be transformed into a conclusion, and that is where algorithms come in. An algorithm is essentially a set of steps that start from an initial situation (the input) and aim for a final goal (the output). In expert systems, these algorithms often take the form of 'if-then' rules. For example, a rule for a robot vacuum cleaner might be: 'If it hits a hard object, it should turn right.' These systems are called 'expert' because their rules are not generated by the machines themselves but are written by humans—specifically, experts in the field. Taking MYCIN as an example again, the rules were not based on mere hunches but on medical doctors' deep knowledge of bacteria and infectious diseases. Thus, expert systems are largely the product of human effort: first, an expert's specialized knowledge is tapped, and then this knowledge is coded into a program by a programmer.

An expert system has the advantage of being transparent. The moment a decision is made, it is not difficult to find out on what that decision was based. One can always fall back on the inputs and the rules. But there are also problems associated with such technology. Capturing and writing the rules takes a lot of time, and when the system enters a new situation for which no rules were written, it does nothing. A solution to these shortcomings comes from that other form of AI: machine learning.

As the name implies, in the case of machine learning, the technology itself is going to figure out how to process an input into output; the AI system is going to learn, on its own, how to link a decision to incoming information. It does this by seeing lots of examples in the learning phase. In other words, the technology is trained with immeasurable quantities of data, such as millions of photos. Unlike an expert system, machine learning works in a data-driven way. In this context, one can think of the spam in your mailbox. It was put there by an AI system that was first trained with a huge number of examples: messages that are undesirable (spam) and messages that we might like to read. Once the system is trained, it must know how to correctly determine incoming messages (input) as either desirable email or as spam. That prediction is the output. If it is able to do that, then it is an example of a smart technology that functions on the basis of a self-learning algorithm, without the help of an expert or programmer. Another example of machine learning is the algorithm of the Swedish music platform Spotify. It keeps track of all the decisions made by its users on the platform and searches for similarities between users' choices. If the algorithm finds them, then users with similar behaviors are grouped together. Now, when someone in that group listens to an album—*Stadium* from 2018 by Eli Keszler for example, a great album by the way—that record also appears to the members of that same cluster as a listening suggestion. Again, incoming data (the set of choices on Spotify) is linked to an output (the suggestion), and, again, no human intervention is required.

In this context, it is important to clarify that machine learning and deep learning are not the same thing (Dignum, 2019). Although all deep learning is a subset of machine learning, not all machine learning involves deep learning. Essentially, deep learning is a specialized branch of machine learning that trains technology to transform data (input) into decisions (output). What sets deep learning apart from other types of machine learning is its simulation of the human brain's structure. The brain consists of neurons—cells that transmit signals to one another. Deep-learning technologies replicate this arrangement using neural networks, which include multiple layers of artificial neurons. Typically, these neural networks have many layers. The term 'deep' in 'deep learning' refers to these multiple layers, underscoring the depth of the neural structures involved.

When people talk about AI, they sometimes just mean expert systems such as MYCIN. Such technology is used for spelling and grammar checking, for example. In other cases, it refers to the combination of expert systems and machine learning—that combination, incidentally, is also the way in which AI will mainly have to run in the future, according to technologists. Usually, however, by 'smart systems' people mean technologies that are based solely on a self-learning algorithm. Examples include Netflix's software that recommends series and movies, and AI systems that help doctors to detect cancer. We follow that use of the term 'AI.' In the chapters that follow, by 'AI' we *never* mean superintelligent technologies or general AI. When we talk about AI, we are *usually* talking about technology with a self-learning algorithm, *sometimes* about machine learning combined with expert systems, and *hardly* ever exclusively about expert systems.

A Philosophical View

This book presents an introduction to philosophy and ethics of technology and AI, and there is a compelling reason for this approach. Designers, vendors, users, and producers of technology and AI are all confronted with critical issues such as sustainability, equality, transparency, privacy, bias, and responsibility. These challenges are highly relevant in our twenty-first century, especially given the impact of both sophisticated (smart) and simpler (dumb) artifacts that we create. Each of these artifacts raises ethical and moral questions, directly tying into ethics or moral philosophy, a crucial branch of philosophy. This connection alone justifies the need for a philosophical lens in discussing technology and AI.

The Essence of Philosophy

So what follows is not an introduction to technology and AI, even though we explain examples of them at length in more than one place. Nor does the book provide an empirical, scientific framework for thinking about smartphones and steam engines. Thus, although we are obviously informed by empirical studies and do not engage in armchair philosophy, we purposefully omit a series of fascinating and relevant questions from these fields of science. From economics: will this or that job disappear? Is there a limit to automation? (Agrawal et al., 2018; Ford, 2015). From psychology: why is it that people use a lot of technology but often react with reticence when a new development is brought to market? What are the political beliefs of the creators of technology and AI? How should we interpret the uncanny valley phenomenon, the fact that people react aversely to robots that are very similar to humans? These and many other questions need to be answered by economists, lawyers, and psychologists. We do not highlight them in this book: not because they would not be challenging or relevant, for they are at least as much so as the philosophical questions we raise (and perhaps even more so), but rather because they are not questions destined for the domain of philosophy and we are familiar only with some of the issues in that domain.

So what does the book offer? What does a *philosophical* view entail? That question suggests that there might be something common to all that we call 'philosophy,' that there is something that all philosophical texts and reasoning share. In addition, that question also seems to probe for something that is unique to philosophy, for a characteristic that we attribute exclusively to philosophy. However, is there something that is characteristic of *all* philosophy and *only* philosophy?

A popular conception of philosophy is that it shines a critical light on things. Large parts of the chapters that follow are indeed an outright attack on deeply held beliefs, but on the other hand, criticism is not the privilege of philosophy. Designers and bakers, garage owners and mathematicians, there is little or no reason to suspect that they would be less critical than philosophers—the opposite is probably true—let alone that they would *not* be critical. Another possibility is to look at the methods that philosophers use when doing philosophy. Later, we set up some thought experiments—unusual imaginary situations—with the goal of testing some of our intuitions about technology; for example, the belief that robots cannot suffer. Now, if one has the impression that this isolates philosophy from other disciplines, that is unjustified. Scientists also make use of such experiments—think of Isaac Newton's thought experiments—and, furthermore, there are philosophers who never use such methods. Another possibility is to look at the kinds of questions that are asked in philosophy. They are not only difficult to answer, but additionally, and above all, they are often very abstract. This is true for numerous questions coming from philosophers, but certainly not for all questions. When an ethicist thinks about euthanasia, vegetarianism, or migration, these are far from easy topics. However, these are certainly not extraordinarily abstract—at least they are less abstract than some questions posed by theoretical physicists.

We could be mistaken, of course, but there is probably no unequivocal answer to the question of what exactly philosophy is and is not, at least insofar as that question refers to a characteristic that characterizes only philosophy and all of philosophy. Put another way: when people say something is 'philosophy,' they can mean different things by that, just as 'technology,' 'AI,' or say, 'sustainability,' are multivalent terms. Some may find this annoying, as we tend to differentiate and neatly pigeonhole things. In that case, let this be a consolation: philosophy is not alone in that regard. There are fields other than philosophy that cannot be lumped together under one heading. And what, for example, would be the property that characterizes all and only sports, war, or religion?

Describing and Judging

In itself, it does not need to be particularly problematic that there is no conceptual unity, that under the rubric of 'philosophy' there are diverse things that sometimes have little to do with each other. It just means that we need to be clear about what we mean by a philosophical view here in this book. It is good, therefore, to begin by explaining exactly what that gaze will be directed at.

People have deep-rooted and widespread beliefs about a broad variety of things. They are ideas that have been around for a while, sometimes for several decades or even centuries, and that recur in different walks of life. Some of these beliefs are noncommittal or uninteresting, whereas others are not without practical consequences. Some examples: people think or thought that they have a soul (and that animals and robots do not), that happiness is the goal of life, that you have to eat meat to be healthy or to live well, that success is exclusively dependent on their own efforts, that there is a world outside of them, that hard work is the gateway to happiness, that capitalism is heaven (or hell), that the earth is flat and at the center of the universe, that enjoying life is important, that they may or may not have been created by God, that they are the crown of creation or uniquely special, that they are free—and a whole host of other things.

This book delves into the deeply held beliefs surrounding technology and AI, perspectives that have evolved over centuries and are shared widely—from tech giants such as Mark Zuckerberg and Google's founders to designers and computer scientists, and extending beyond the tech industry to politicians, philosophers, scientists, and opinion leaders. This book specifically focuses on three prevalent assertions: technology is neutral, AI is a disruptive technology, and technology must be understood in terms of determinism. Although there are other views we could explore—such as technology as applied science, which we touch upon in Chap. 3—we have chosen to focus on these three beliefs for several reasons. First, they are the most widely recognized. Furthermore, these beliefs are not merely theoretical; they resonate with real-world implications and are deeply entrenched in practical outcomes, often with significant consequences.

Thus, this book is about the assumed neutral, disruptive, and determinant nature of technology. It draws attention to the narratives that are routinely peddled about technology and AI in workshops and advertisements, the theses on which sometimes undesirable tech practices are founded, the familiar talk of Silicon Valley, the creeds of technophiles and pessimists, the theses of cyberutopians and techno-alarmists, the way in which technology is talked about today and has been talked about for a long time by policymakers and advertisers, the ingrained beliefs that we, consumers and users, have about the things we carry with us all the time—smart technology and functional artifacts that are not AI. Think of the popular stories about self-driving cars being the future of transportation and smartphones destroying human connection, among others, to see what is at stake in this book.

Chapter 1 zooms in on the supposed neutrality of technology, and is about technology in general, not just AI. Among other things, we focus on a moral value such as fairness and on whether technology is laden with such a value. To do so, we invoke bridges designed with racist intentions, discriminatory algorithms, drilling machines, *Temptation Island*, and gender-neutral video games. The second section, on AI as a disruptive technology, focuses purely on smart technology, paying particular attention to the following ethical issues: privacy, bias, moral responsibility, ecology, transparency, abuse, and security. In the third chapter, we highlight the thesis that one must understand technology—and by this we are not only referring to AI in that chapter—in terms of determinism. Is technology a product of society or is society driven by the work of data scientists, engineers, and computer scientists? We visit numerous examples: Starbucks' scheduling software, nuclear power plants, bicycles, social media, the QWERTY keyboard, and the steam engine.

In each of the three chapters of this book, which are structured to stand alone, we focus primarily on two main tasks, although three areas are covered. First and foremost, we clarify the concepts being discussed. The assertions about technology that we examine—such as the neutrality of technology or its deterministic impact on society—have been widely accepted for years and are commonly echoed. However, when pressed for specifics about what these assertions truly entail, the explanations often dissolve into vague generalities. This ambiguity prompts our first objective: to offer a crystal-clear exploration of these claims. What exactly do terms such as 'neutrality' and 'determination' mean in the context of technology and AI? What does responsibility entail in this framework? Are there multiple interpretations, and if so, how are they interconnected or distinct? By pursuing precision and clarity, our goal is to cut through the fog of ambiguity, not by layering vague ideas on top of each other but by dissecting and differentiating the claims and concepts at hand. This approach will help to lay a solid foundation for understanding the intricate dynamics of technology in society.

Second, we undertake a thorough evaluation. Once we have established a clear understanding of the assertions under discussion, it is time to assess their validity. What are the arguments for and against the idea that technology is neutral? Which arguments are the most convincing? Is there substantial support for the claim that AI is a disruptive technology, or are there compelling reasons to question or even reject this notion? The central question we are addressing, is whether these popular, entrenched narratives hold true. Here, we offer a preliminary answer: some technology may indeed appear neutral, but this is not necessarily the case—many technologies are imbued with values. From an ethical standpoint, we see no substantial reason to categorize AI as inherently disruptive. Furthermore, the arguments linking technology with determinism are significantly flawed; indeed, we contend that all these propositions are incorrect. In each chapter, we delve deeper into why we hold these views, providing detailed analysis and justification for each position.

Last, in each section, beyond understanding and evaluating, we address whether our arguments have merely theoretical significance, or whether they also bear practical importance. For instance, is it merely an academic exercise to determine whether technology is or is not neutral, or does this knowledge have real-world applications? Take the concept of technology's neutrality, or lack thereof. Understanding that technology often carries embedded values is crucial, particularly because these values can be hidden. This awareness can prevent us from unwittingly adopting technology that espouses values with which we fundamentally disagree. Knowing this can influence our decisions about which technologies we choose to use or avoid. We explore this aspect further, along with other issues, in the detailed discussions to follow in each chapter.

The core of the philosophical approach of this book is to clarify and challenge self-evident truths. It is also what other philosophers throughout history have seen as their main task and is in line with what Friedrich Nietzsche meant by 'philosophizing with a hammer.' The choice of such an approach does have the consequence that some other possible perspectives are not central or hardly play a role—for example, a historical perspective. Not that this would be impossible: although philosophical reflection on technology and AI is not old in comparison with the long history of philosophy, it has still existed for about a century and a half. Philosophical thinking about technology got off the ground in the second half of the nineteenth century—in particular, once Ernst Kapp had published his *Grundlinien einer Philosophie der Technik* in 1877. Moreover, numerous philosophical studies of technology have appeared in the last hundred years. We are thinking mainly of the work of Martin Heidegger and Karl Jaspers, and the studies of Bernard Stiegler or

Gilbert Simondon. The format of our reflection, however, implies that the pages that follow do not sketch a historical overview of the philosophy of technology, and that we do not present an in-depth study of the work of those authors. Such philosophy is, of course, fascinating, and even though we devote more than one page to Heidegger and in some sections connect to the work of a number of philosophers of technology, a history of philosophy of technology, brief or otherwise, or an analysis of a philosopher's discourse on technology and AI is not necessary here.

Our approach ensures that this book does not squarely align with either staunch optimism or deep-seated pessimism about technology, although it would not be surprising if it seemed to lean one way or the other at times. In many discussions, particularly those led by entrepreneurs or technologists, there is a tendency to extol the limitless potentials of technology. Conversely, studies emerging from the humanities, especially those authored by philosophers, often adopt a distinctly alarmist tone. Throughout this book, we do indeed voice significant criticism, for instance, concerning the ecological impacts of technology and AI usage. Simultaneously, we acknowledge and praise the technological strides made in areas such as medicine. By steering clear of the biases that often color the views of technophiles and technophobic alarmists, this book maintains a balanced perspective. The primary focus is on describing and evaluating the issues at hand. From this foundation, both critical insights and optimistic viewpoints naturally emerge.

Chapter 1
The Neutrality of Technology

> *Technology is usually fairly neutral. It's like a hammer, which can be used to build a house or to destroy someone's home. The hammer doesn't care. It is almost always up to us to determine whether the technology is good or bad.*
>
> Noam Chomsky

Texas, spring of 2018. A horrific shooting takes place at a high school. In response to this terrible tragedy, Oliver North, president of the National Rifle Association (NRA) says, "Guns don't kill people, Ritalin kills people!" In the aftermath of the shootings in Ohio and (again) in Texas, a year later, the former President of the USA Donald Trump says something similar: "Mental illness and hatred pull the trigger, not the gun."

Both statements riff on the slogan 'Guns don't kill people, people kill people!' first used about a century ago. Its origins are deeply rooted in the tumultuous history of early twentieth-century America. During the 1920s, the USA experienced a significant surge in violence, much of it fatal and frequently involving gang confrontations in bars, where disputes were often resolved with gunfire. This wave of violence led to a notable reaction in 1922 when the American Bar Association called for the cessation of gun manufacturing for civilians, a stance that starkly contrasted with the rights granted under the Second Amendment, which had been in place since 1791.

This call from the American Bar Association resonated with many in the public who were concerned about the level of violence, but it was met with strong opposition from the gun industry. Gun manufacturers were particularly alarmed by this challenge to their business and to what many of them considered a constitutional right. Their discontent culminated in various forms of protest, the most notable being an article published in 1927 in *The Manufacturer*, a trade magazine for American factory owners. The article famously stated, 'Guns don't kill people, people kill people!,' marking the inception of what would become a longstanding slogan for advocates of gun ownership.

The enduring power of this phrase is evident in its later adoption by the NRA, which further popularized it in the latter half of the century by distributing bumper stickers featuring the slogan. This simple yet contentious statement continues to encapsulate the central argument in debates over gun control and rights.

The slogan 'Guns don't kill people, people kill people' has transcended its original context related to firearms and has been co-opted into broader discussions about technology. This adaptation reflects the ongoing debate surrounding the neutrality of technology, a concept we refer to as 'the neutrality thesis.' This thesis proposes that technology itself is neutral, devoid of inherent values or biases, and that any impact it has stems from how people choose to use it. The argument is not unlike the contention in the gun debate that firearms, as objects, do not cause harm independently; rather, it is the individuals wielding them who bear responsibility for their actions. Similarly, the neutrality thesis suggests that technology, like a tool, does not contain ethical or moral weight by itself.

However, the neutrality of technology is a contested notion. Just as there are clear cases where staying neutral is impossible, or even undesirable, such as parents making choices favoring their children in emergencies, there are also scenarios in technology where built-in biases or designed functionalities challenge the idea of inherent neutrality. Thus, although the slogan from the gun debate has been repurposed to support the neutrality thesis in technology discussions, it also opens up the floor to scrutinizing whether technology can truly be neutral or if, like all human creations, it carries with it the biases and intentions of its creators.

Those who claim that cell phones and apps, for example, are neutral, are not making a new or radical observation. The neutrality thesis is at least as old as (Western) philosophy itself—it already existed in ancient Greece. It was the Stoics who, around the fourth century BC, said that technological artifacts are neutral.

The neutrality thesis not only goes way back, but it is also widespread. This is also what one can read on the opening page of Heidegger's infamous 1954 essay *Die Frage nach der Technik*, where it is written that we are delivered over to technology in the worst possible way when we regard it as something neutral (Heidegger, 1977). Heidegger wrote his text in the middle of the last century. If we look at our era, where do we still find the idea that technology is neutral?

Philosopher Joseph Pitt defends the famous thesis in his 2014 text titled '"Guns don't kill, people kill"; values in and/or around technologies.' In the book *Hello World—How to be Human in the Age of the Machine* by mathematician and radio producer Hannah Fry, one can read the assertion that no object or algorithm is ever either good or evil in itself, that it is only about the technologies that are used (Fry, 2018).Then again, linguist and political activist Noam Chomsky stated unequivocally in 2014: "Technology is usually fairly neutral. It's like a hammer, which can be used to build a house or to destroy someone's home. The hammer doesn't care. It is almost always up to us to determine whether the technology is good or bad." (as quoted in Veletsianos, 2014). And finally, the neutrality thesis was also present in the background of Mark Zuckerberg's hearing in the US Senate in April 2018. When Senator Ted Cruz took the floor and addressed the Facebook founder and CEO, he asked Zuckerberg if Facebook is a neutral public forum. Zuckerberg replied that Facebook is a platform for all ideas.

So even today, that technology is neutral is a widely held proposition. But is this well-known and oft-quoted idea about technology right or wrong? In the quotation above, Heidegger not only states that the neutrality thesis is very popular, but he also states unequivocally that it contains multiple flaws. But what exactly? Is technology not neutral? Who is right, Heidegger or Zuckerberg?

At first glance, Chomsky's assertion that tools like hammers are neutral seems straightforward. However, numerous examples challenge this neutrality thesis. Consider the use of artificial intelligence (AI) in courts to predict recidivism risks. In 2016, the COMPAS algorithm, which stands for 'Correctional Offender Management Profiling for Alternative Sanctions,' came under scrutiny by *ProPublica*. Although the algorithm's designers did not intentionally include ethnicity as a factor, their analysis revealed a bias: darker-skinned individuals were more likely to be incorrectly assessed as high risk (false positives), whereas lighter-skinned individuals were more often mistakenly judged as low risk (false negatives). For instance, a woman of color who stole a bicycle was deemed a higher risk than a white man with a history of armed robbery. This example suggests that even designs intended to be neutral can, upon closer inspection, perpetuate inequality based on irrelevant characteristics such as skin color—an issue we refer to as 'algoracism.' Although unintentional, such biases in AI can be as damaging as CO_2 emissions are to the environment, a topic we explore further in the Chap. 2.

Are technologies neutral? Or are the algorithms with which those technologies are equipped 'weapons of mass destruction' (O'Neil, 2016)? Stones are stones. They are 'closed,' *en-soi*, according to Jean-Paul Sartre. But is the same true of technology? Is there 'an opening' in these abrasive, squeaky, creaking, steely things, an opening through which values, norms, ideologies, or stereotypes can penetrate?

In the following pages, we reflect on the neutrality thesis. As announced in the Introduction, here too we roughly follow three strands. First, we try to understand. What exactly does the neutrality thesis entail, and what does it not entail? To do this, we need to zoom in on the term 'value.' What do we mean when we say that, for example, privacy and fairness are values? Second, we evaluate. What are the advantages and disadvantages of the neutrality thesis? We argue that the neutrality thesis is false, but that there is still some truth in it: some technology is neutral, and some is not. Third, we ask whether our analysis has practical implications. Are the forthcoming paragraphs relevant to engineers, computer scientists, and AI developers? We answer this question toward the end, in the affirmative. What we assert has implications for our moral relationship with technologies and technologists, with things and people, dead matter and organisms.

Having and Being

In a sense, one is stating the obvious when saying that technology is not neutral. After all, it is obvious to everyone that technological tools do not exist for just anything; they are made with a specific purpose in mind. There is therefore always already a choice underlining the very creation of the artifact, a choice made by the designers to realize some particular goal by means of the technology. Everyone knows this, and so no one denies that technology is not neutral in this sense. So is that the end of the neutrality thesis?

One can roughly distinguish a broad and a narrow interpretation of the thesis. The broad interpretation says that no technology is laden with norms, perspectives, and views, that these things are not part of the technology. If that were the case, it would mean that there is a high probability that the use of a technology will express certain world views or stereotypes, or that it will be in line with some norms or values. But, according to the broad interpretation of the neutrality thesis, this is not the case; technology is not tied to a particular view or ideology.

We call the other interpretation 'narrow' because it deals *exclusively* with values. It says that technology in itself is independent of values, that it is not value laden. It is, rather, value neutral or value free. This means that a design as such gives no reason to believe that the state of affairs that will be realized by using the technology will be in line with a value like autonomy or sustainability. What results from the use of an app or bridge may be in line with a value, but that is not what you expect based solely on the construction of the app or bridge. The alignment results from the use, and not the design, of the technology. At least that is what one believes when the narrow interpretation of the neutrality thesis is defended.

In this chapter, we zoom in exclusively on the latter interpretation. There are a number of reasons for this. First, the interpretation that deals with values is the best known and the most frequently cited. In addition, unlike the broad interpretation, it is about ethics, and that is precisely one of the areas of special interest to philosophy. Finally, the broad interpretation is not very credible. Technology is clearly not necessarily separate from, for example, stereotypes or ideologies. This becomes apparent when you take a closer look at a number of examples. We start with a now-familiar device.

Decolonizing Technology

In 2017, a tech industry worker from Nigeria called Chukwuemeka Afigbo tweeted a video with the following caption underneath: "If you have ever had a problem grasping the importance of diversity in tech and its impact on society, watch this video." The video shows a device that dispenses soap when you hold your hand under it. First, the hand of a person with white skin appears on the screen, and the device squirts soap on it. But when Afigbo holds his hand under the device, the device does not respond. Finally, when he holds a white cloth under it, the soap device does function properly again. Does this mean that the designers had racist motives? Of course not. But the video does show that technologies are not detached from the designer's perspective or worldview, that they can be driven by a white perspective, for example. And in this case, it is a rather uncomfortable truth.

Another argument against the broad interpretation of the neutrality thesis comes from the field of disability studies. There are smart assistants that do not recognize the speech of people with amyotrophic lateral sclerosis (ALS); some AI systems for facial recognition do not work for people with disabilities. Another example comes from the company HireVue, which sells smart systems to companies in order to select the best candidates for a job. An interview with a candidate first takes place and then AI analyzes the conversation, taking into account the tone of voice, facial movements, and patterns in speech, among other things. Based on the results of that analysis, the technology selects a candidate. Although such a system is good in terms of efficiency, it is less good in terms of ethics. A report examining HireVue's technology states that the technology discriminates against many people with disabilities that significantly affect facial expression—one can think about speech stroke survivorship (Engler, 2019; Whitakker et al., 2019). Despite the fact that there is clearly ableism at work here, this does not necessarily mean that the designers have bad intentions. However, one can conclude from the examples given that some technologies are designed from a well-defined frame of reference—that of people without too many deviations or limitations—and that these technologies are therefore not independent of such a frame.

In the Introduction, we pointed out that most smartphones are made with the average length of a man's hand in mind. By extension, so is the voice recognition software used for phones. Many mobile phones from Samsung are not equipped with such technology, but should they be, there is a real chance that it would work better for recognizing a man's voice than a woman's. In a way, that is surprising, because we know from research that it is easier to understand a woman's voice than a man's. On the other hand, it is not very surprising. The technology is designed using datasets of recorded voices. And what does it show? The majority of voices come from men. So based on this example, one can say that technology can stem from a male view of the world, and thus that a certain masculine view can be attached to that technology (Perez, 2019).

Decades ago, razor-blade manufacturers made a distinct separation between men's and women's razors. Women's razors were designed to be nonserviceable, lacking the ability to be opened for repairs. Conversely, men's razors not only opened but often came with a leaflet and sometimes brushes for maintenance. This design choice reflects the underlying, traditional gender stereotype that women are either incapable of or uninterested in repairing such items, suggesting that such tasks are suited only for 'real men.' This example demonstrates that even basic technologies can carry ingrained cultural biases, challenging the notion that technology is always neutral and devoid of nontechnological issues such as sexism.

The Grandfather's Watch

So in this chapter we focus not on the broad but on the narrow interpretation of the neutrality thesis. We already mentioned that it means that technology is independent of value. But what exactly does that mean? Now, the problem is not so much what we should understand by 'technology,' the difficulty has more to do with 'value.'

For although the meaning of that term is more or less clear at first sight, a closer look reveals that it is not crystal clear what is to be understood by it. What, for example, was referred to when it was said in the debate about the corona tracking app that values are also at stake, a value such as privacy or fairness, for example? Let us therefore outline the contours of the meaning of that term.

When you want to know what 'value' means, it is appropriate to keep in mind that this word, like many other words, has multiple meanings (Van de Poel & Kroes, 2014). We generally use the term in two ways in everyday life. You can say that something *is* a value or that something *has* value. See some examples here. On the one hand, we have sentences such as 'The AI system for recruitment has value' and 'His late grandfather's watch is valuable to Tom'; on the other hand, it can be said that psychiatrists always keep a value such as integrity in mind, or that autonomy has been a central value in bioethics for several decades. Here, one immediately sees that 'value' takes on a somewhat different meaning in the first two sentences than it does afterwards. What exactly is the difference?

Look at the sentences about the AI system and the grandparent's watch. They claim that these technologies have value. When the word 'value' is used in this sentence, it means one of two things: either that something has instrumental value, or that something has non-instrumental value, and thus has value in and of itself.

To say that something has instrumental value means that it is an instrument that serves a goal, that it plays a role in the realization of a purpose. This can be said about many things, such as artifacts and deeds, for example. A hammer and telephone have such value, because one can use them, respectively, to hammer a nail into the wall and to make a phone call; studying has value, because it helps to pass an exam. In addition, one can attribute instrumental value to non-artificial things. A tree trunk can be used to sail down a flowing river; one can make use of a stone to make fire.

One can attribute value to something beyond its practical use, much like the sentimental worth of a grandfather's watch. This suggests that objects can possess intrinsic value, independent of their utility. Although usefulness often qualifies something as valuable, it is not a necessary condition—items without practical utility can still be valuable. When we recognize such non-instrumental value, we typically express a positive judgment, indicating a favorable view of the item. This judgment might be based on aesthetic appreciation, such as when something is deemed beautiful, or an emotional connection, such as the watch reminding Tom of his grandfather and their cherished relationship. Ultimately, recognizing an object's value naturally leads to treating it with care, protection, and reverence.

Moral and Not Moral

But there is a second meaning of 'value,' which comes to light in the other sentences that we presented, the ones that speak of a value such as autonomy or respect. In that case we are not talking about something that *has* a value, but about something that *is* a value. We do not normally mean that autonomy has value, but that it is a value. If uses 'value' in this sense, one is not expressing a judgment (as with 'having value') but referring to a certain state. For example, we say that freedom is a value. One is then referring to a particular state of affairs, and more specifically, one is probably referring either to the state in which a person or group is not coerced or limited by others, or to the state in which one has the possibility to choose from several options. It is useful to explore these issues in more detail.

Numerous values are aligned with the concept of freedom, including respect and sustainability. They share a moral dimension. They all concern the ideal or desirable interactions among people, between humans and animals, between humans and nature, and between humans and technology. However, there are also values that are amoral in nature. For example, in the art world, values such as beauty or elegance often, but not always, take precedence. Similarly, in the field of engineering, efficiency is a key value. Although beauty and efficiency are values akin to sustainability, they differ in that they do not contain moral implications. Both represent values of a morally neutral nature.

We have yet to refine the similarity between, say, beauty and sustainability. After all, values are not mere states of affairs. They are states of affairs that we consider to be important. It is contradictory to say that something is a value on the one hand and to think that it does not matter on the other. In fact, with 'value' we always refer to a state that we think is so important that it *must* exist or be taken into account. For example, if one says that autonomy in health care is a value, one does not mean that people actually have control over their own lives. Nor do you mean exclusively that it is important for people to be allowed to shape their own lives. No: those who say that autonomy is a value in the health sector are doing more than describing it or pointing out its importance. They are saying that people should be treated as autonomous human beings in the health sector.

An even more precise description of values is that they are often of a non-instrumental nature. In other words, they refer to things that we consider important in themselves. This means that one strives for, say, beauty and autonomy, not because they have something else in mind, but because of the values themselves. Therefore, values themselves constitute a goal, and not a means to another goal. Perhaps the best-known example of this is happiness. This is without doubt a value, but moreover a value that is an end in itself. You do not normally say that you want to be happy in order to achieve some other goal. No, you want to be happy, period.

Does that mean that values are necessarily non-instrumental in nature? No. Although it would be peculiar to claim that one wants to become happy in order to achieve something else by doing so, values can have an instrumental character in addition to being non-instrumental. By this we mean that we pursue some values as a means of achieving something else. This may sound a little strange at first glance, but it nevertheless corresponds to the role that values play in everyday life. Take privacy, for example. Of course, we find it reprehensible when our personal data are up for grabs without our consent, regardless of the consequences. We want privacy for its own sake. But protecting privacy also serves to shield citizens from abuse by companies and governments. Privacy is also a tool to achieve other goals.

Finally, values are not necessarily universal, in the sense that they are not shared by everyone and that not everyone values them to the same degree. Take the discussion about the use of an app designed to limit the spread of SARS-CoV-2. In Europe, a good portion of the debate was about the risk that the app would not respect the user's privacy. In Asia, such issues were much less at play. The differences can be very large and lead to profound disagreements. Most people think inclusiveness is important, for example, but (unfortunately) some people find it reprehensible. They push forward not inclusiveness but its opposite, segregation, as the condition to try to achieve. Segregation is a value to them, which we—and many others with us—find abhorrent and repugnant.

To conclude, we'll list the most important distinctions. The term 'value' can be used in two ways. We can say that something, sustainability, for example, is a value, and we can say that something, a watch or a house, has value. In each case we can distinguish an instrumental and non-instrumental interpretation. A value such as happiness is important in itself, whereas a value such as privacy also serves other purposes—financial security for example. And as for 'having value,' a watch helps you not to lose track of time, and can also have value in itself, because it reminds you of your grandfather.

Technology and Ethics

As previously discussed, the neutrality thesis in its narrower form posits that technology is value neutral. This is a claim often made by those deeply involved in the development and spread of technology and AI. Two key terms here are 'technology' and 'value.' The term 'technology' is commonly used in two senses: it can refer to artifacts with a functional purpose (such as a laptop) or to processes of technological manipulation (as in 'biotechnology'). In this text, we primarily discuss the former, as outlined in the Introduction.

We are now equipped to explore what 'value' means in this context. We demonstrate shortly that according to the neutrality thesis, technology itself does not inherently embody moral values, such as privacy or sustainability.

Crying About a Robot

From October 2017 to April 2018, the 'Hello Robot' exhibition at the Design Museum in Ghent, Belgium, explored the interaction between humans and technology. The exhibit included a scene from the 2013 movie *Her*, where Joaquin Phoenix's character falls in love with an operating system. Additionally, those who grew up in the 1990s experienced a nostalgic moment with the display of a Tamagotchi, the egg-shaped virtual pet. Also among the exhibits was a robot encased in glass that, despite its unattractive, old-fashioned vacuum cleaner appearance, was a highlight. This was due to it being R2-D2, a beloved droid from the *Star Wars* saga. The special handling of the robot by a confidant of *Star Wars* creator George Lucas added to its allure.

In November 2005, in Fallujah, Iraq, US Army soldiers organized a funeral for Boomer, a MARCbot robot used to dismantle explosives, which had been destroyed by enemy forces. The soldiers held a sentimental ceremony complete with military honors: two medals, the Purple Heart and the Bronze Star Medal, and 21 bullets were placed in Boomer's casket to honor its service (Garber, 2013).

These stories illustrate that technology can hold intrinsic value beyond its practical use. Boomer, though destroyed and no longer functional, was not considered worthless by the soldiers who had grown attached to it, attributing emotional value to the machine. Similarly, R2-D2, no longer active on movie sets, still drew significant interest at the exhibition, indicating that it possesses symbolic value—stemming from its iconic role in the *Star Wars* films and its physical presence during filming.

Should the neutrality thesis mean that technology has no value separate from utility and service, the examples just given would undermine it. However, no one is arguing that technology per se cannot have non-instrumental value. The mere fact that technology is usually material is enough to realize that technology can be symbolically or emotionally laden with value. Because of that tangible nature, you can inherit it from someone dear to you or you can become attached to it in an emotional sense. So anyone who claims that technology is value free does not mean that technology cannot have value in itself, and therefore cannot answer you by referring to things such as R2-D2 or Boomer.

One may not conclude on this basis that the neutrality thesis is about that other value, about instrumental value. The reason is that there is an inextricable connection between such a kind of value and technology. The connection is forged by the property that by definition belongs to technology: functionality. All technology, we saw earlier, necessarily has a function, in the sense that the technology is made to realize a purpose. To put it another way, all artifacts must be purposeful in order for you to see them as technology. If something is man-made but has no function, you cannot say it is technology. Now, if we assume that all technology works properly, then this reformulation makes it clear that technology has instrumental value in all cases. Indeed, if you cannot say an artifact is technology if it serves no purpose, then it follows that all technology necessarily has instrumental value. In short, it is impossible for a technology to do what it was designed to do and still not have instrumental value. This inevitable link between technology and instrumental value is why no one argues that technology is free of instrumental value.

Without Moral Value

We can now begin to see what the neutrality thesis means. If it is self-evident that technology can have value in itself and that it necessarily has instrumental value, then it must be the case that the neutrality thesis has everything to do with, for example, privacy and fairness, things that we consider *to be* moral values. To get a good idea of this, we shall dwell a bit more on the instrumental value of technology.

One can use technology for different kinds of purposes. A drill that you can use to assemble a cabinet, an exoskeleton that you can use in tough work situations to reduce the strain on your body. In addition, values can also be goals—although not all goals are values. You can use technology to strive to achieve a state that you feel is so important that it must be realized. Of course, life is no walk in the park, and of course we are not naive techno-optimists, but the number of technologies that contribute to the realization of values is almost endless. Traffic lights lead to increased safety, solar panels have a positive effect on sustainability, wheelchairs make society more inclusive, and medical technology helps to prevent, detect, and cure diseases. Or take the use of algorithms in the administration of justice. Although we mentioned earlier that this is not without its problems, it can also ensure more equal treatment of cases. If you leave legal judgments solely to judges, then there is a real chance that two identical situations will be judged differently, depending on, for example, the time of day at which the judgment is rendered. We know that judges are more harsh when they are hungry, just before their lunch breaks. Algorithms, on the other hand, are at least in this respect a good thing in that they are not bothered by circumstances such as hunger or fatigue.

In the examples just given, technology is linked to a value such as privacy or fairness. It is now important to know that this link is made in a well-defined context. The AI systems, wheelchairs, and medical devices we just mentioned are linked to values when they are used. The linkage occurs when one takes the technology in one's hand, engages with it, and deploys it for a purpose. The neutrality thesis is also about values such as privacy and fairness—we already knew that—but not in the first place about the technology's phase of use. It is about technology in itself. More specifically, the thesis is that when you consider technology, no values attach or cling to it. Values can be influenced by technology, for better or for worse, but technology itself is independent of values. Technology is neutral: not tied to this or that value, not for or against any value. In other words, if you know a technology, then, according to the neutrality thesis, you have no reason to suspect that the use of the technology will be aligned with a value. Sure, it is quite possible that you might use a technology in a way that realizes a value, but that does not follow from the construction of the technology, from the designer's intention. The value is not part of the technology, the technology is not intimately intertwined with this or that value. See here the explanation of the popular neutrality thesis, at least of the narrow interpretation.

As that thesis is the central theme of this chapter, it is important that there be no misunderstanding. Let us therefore make three more points. First, the claim is not merely that all technology is value free. Indeed, that could still mean that although it is theoretically possible for technology to be value laden, it is de facto not. But the claim is stronger. The neutrality thesis holds that technology is not value laden because it cannot be value laden. In other words, technology is by definition value free, according to someone such as Chomsky or Pitt.

Second, it is important to note the terms often used to explain the neutrality thesis: 'sticking,' 'loaded,' 'hanging on.' It goes without saying that these terms are used here exclusively in a figurative sense. No one, not even those who reject the neutrality thesis, is claiming that values are literally in technology and actually attached to it. Indeed, by the term 'value' we are referring to a state of affairs in reality outside of technology. Justice and respect, for example, have to do with the relationship between persons, just as sustainability refers to the relationship between organisms; these values are not part of technology, they are separate from it. So what does the term 'value ladenness' refer to? The figurative interpretation of the term is that the design of a technology gives reason to believe that the use of technology will be in line with some value; based on the knowledge of a technology, you can expect that the technology will not harm a value; the design of the technology makes it very likely that if used properly, the technology will realize a value. According to the neutrality thesis, this is not the case with just any technology. Later, we will examine whether this is true.

Third, keep in mind that the neutrality thesis is not about just any value. In a sense, this is obvious, because who would argue that it is impossible for a value such as accuracy or efficiency to attach to technology? The narrow interpretation of the claim that technology is neutral is about a well-defined kind of value, namely values with a moral character, values that are not morally neutral. Thus, anyone who believes that technology is not laden with value believes that technology does not have values such as privacy, sustainability, or fairness attached to it. In short, the neutrality thesis refers to values that have to do with the good life, and not about, say, scientific or technological values. The point is that there are no ethics contained within things.

Capitalism and Religion

We are almost ready for an evaluation of the neutrality thesis. Is the age-old and widespread claim that technology is neutral true? What are the arguments for and against it? The foregoing teaches us that there are certainly two things we cannot use as arguments against that thesis: theories of the causal connection between technology on the one hand and things that are not technology on the other; examples of technologies that are not neutral in the broad sense of the word. It is useful to briefly draw attention to this now.

In *Platform Capitalism* (Srnicek, 2017), the following link between capitalism and technology is outlined. In a pre-capitalist society, there was no wide gap between people and their means of making a living. People took care of their own food and maintenance. That changed with capitalism. We, humans, are no longer the producers of our own food and now depend on others to make things for us to stay alive. We have to go into the marketplace to make money, money that enables us to buy the necessities of life produced by others. The problem, however, is that in the marketplace there are many people providing the exact same services or products. Thus, there is stiff competition that threatens to make you as a player lose out to others who have cheaper products or services to offer. This struggle drives market players to devise all kinds of methods to bring down labor costs and to push up productivity. Child labor, underpayment, night work, and the relocation of the production process to low-wage countries can be understood from this perspective. The same applies to technology. Think of the assembly line introduced before World War I by Henry Ford. This technology was introduced to increase production capacity. Or take planning software, which also returns in the Chap. 3. Today, in coffee bars and restaurants, the weekly rosters for staff are often compiled by AI systems instead of humans. This technology first analyzes the comings and goings of customers in the past and the corresponding revenues. Based on this, it estimates which moments in the future will be the busiest and the quietest. The goal is an efficient use of staff. There must be enough workers when they are really needed, there must not be too many staff when they are not. In short, technology too is the result of a capitalist logic—or *some* technology at least, because not all technology is embedded in a system aimed at stacking profits.

Another theory concerns not the cause of technology but its effects, especially on religion. The theory goes like this: the use of technology has enabled people to better and more quickly prevent and overcome poverty, hunger, and disease. This has led to a decrease in the need to invoke supernatural powers, religious objects and sacred sites. Technology thus leads not only to profound changes in the labor market, among other things, but also, it is argued, to the desacralization of society.

We do not want to address the question of whether these explanations are all equally correct or plausible right now. We only mention them because we want to draw attention to the difference between this type of theory and the neutrality thesis. Although the neutrality thesis is about technology itself, the theories outlined above are not. Those are about the causal link between technology and something else, such as capitalism and religion. They do not focus on technology, but on the cause (capitalism) or the effect (desacralization) of technology. Thus, defending the proposition that technology is value neutral does not preclude, at least in principle, one from also claiming that technology is a product of capitalism or that more technology leads to less religion. Indeed, the nature of the two statements is different. Certainly, those who suggest that technology is a product of capitalism or that it harnesses the power of religion may also think that this bears consequences for the conception of technology. And yes, it is possible that the causal link between technology and, say, capitalism has an effect on the moral neutrality of technology. But because we are talking about two different kinds of claims, the claim that technology is a child of capitalism or that it is the engine behind the decline of religion is not in itself an argument against the neutrality thesis. This is so for these two claims, but also for other theories that are not about capitalism and religion, but are about a cause-and-effect relationship of which technology is a part.

We also want to emphasize again that our story is not about the broad interpretation of the neutrality thesis, which focuses on perspectives and ideologies, among other things. We focus on the narrow interpretation, about technology and *moral* values. Thus, we cannot use the following case study to critique that interpretation.

Although a few years back SEAT designed a car for women, most passenger cars have always been designed with a man's physique in mind (Gabriels, 2020). This may seem fairly innocuous at a first glance, yet it bears significant consequences. For example, because the seats are made with the average weight of men in mind and because women generally weigh less, the seats often do not provide enough protection for women. On average, women are also more likely to suffer injuries in head-on collisions and whiplash in rear-end collisions. And that has a lot to do with the fact that cars are designed with a male perspective in mind.

Although there is little reason to suspect that car designers have deliberate sexist intentions, these technologies are not neutral. They are designed from a well-defined frame of reference: that of men. Yet one cannot use this case as an argument against the neutrality thesis, at least not against its narrow interpretation. The latter holds that technology is not laden with moral values, although the given case is not about moral values but about a perspective from which technology has been developed. Of course, cars are built with a particular frame of reference in mind (the male gaze); but frames are not moral values. Passenger cars cannot therefore be a reason to reject the neutrality thesis, just as the Nigerian man Afigbo's tweet about the 'white' soap dispenser is not a counter-argument, and just as you cannot use the fact that scissors are usually designed for right-handed people to attack the narrow meaning of the neutrality thesis.

Technology Is Neutral

It is time to change the perspective. We are exchanging a descriptive view for a more evaluative approach. That technology is neutral is a common assertion—but is it true? Are there reasons to agree with it? In the following pages, we provide two well-known arguments for the neutrality thesis: the visibility argument and the dual-use argument. However, we begin by pointing out something recognizable that is neither a technology nor value laden. This is not an argument for the neutrality thesis, but it does show the following. If it is true that technology is value neutral, then technology is an extension of other things that have nothing to do with technology; if it turns out that the neutrality thesis is false, then that claim is nevertheless not complete nonsense either, as there are things other than technology to which no moral values are attached.

Watching Temptation Island

Many things we do are value laden, morally or otherwise. Parents are responsible for the well-being of their children, nongovernmental organizations strive for a more just society, and doctors' actions are attuned to the value of health. But do actions necessarily have values attached to them?

Suppose you watch an episode of the television show *Temptation Island* to relax. If relaxation is a value for you, you may decide that watching the television program is laden with a value. But you can also watch for a reason other than to relax. For example, you are sitting on the couch working on your laptop at night and you decide to watch, albeit only semi-attentively. Watching the program does not relax you; on the contrary, the participants annoy you and you become agitated. Yet the urge to watch is too strong, which has a lot to do with your desire for thrills. You watch, simply to see if any new perils or events will occur. Will Simone give Arda another chance? Does Simone really find Zach to be immature? Although sensation is great fun for quite a few people, watching *Temptation Island* is still not value laden for many. Reason? At the beginning of this chapter, we saw that we consider values so important that we should pursue them and take them into account. However, few—and perhaps none—speak of sensation in those terms. Who gives something such as sensation so much weight that it has an obligatory character? A great deal of what we do is aimed toward a goal. But not everything, not every goal, is a value.

Can we not say that sensation is only the first goal and well-being is the ultimate goal? Perhaps we watch *Temptation Island* because of the sensation it gives us, but sensation is only relevant because it contributes to our well-being. In that case, watching television would be indirectly laden with value, because watching only relates to the value of well-being through sensation. However, it is by no means the case for everyone that watching is value laden in this sense. Why not? The answer goes along the lines of what we just explained: as values are by definition important, something can only contribute to a value if it itself carries some weight. You cannot claim that something is good for, say, your well-being, if it in itself is hardly or not at all important. Because we find sensation meaningless by and large, we also do not say that it is good for our well-being. Note that it is not excluded that some people contradict our reasoning, because for them sensation does have a value or contributes to their well-being. However, that does not take away from the fact that many people, and perhaps even the vast majority, think differently. That in itself is sufficient to link watching *Temptation Island* to the neutrality thesis.

Everything of Value Is Visible

The question now is how convincing the neutrality thesis actually is. Is technology truly value neutral? The most commonly cited argument in favor of this view is the dual-use argument, whereas a somewhat lesser-known argument is the visibility argument. What do these arguments entail?

The visibility argument is split into two parts. The first asserts that for technology to be considered value laden, a moral value must be discernible within it. If you cannot visually identify a moral value embedded in a technology, then it cannot be said to carry one. The second part contends that it is impossible to physically pinpoint where moral value resides within a device. You might see its form and materials, but you cannot point to a specific part that holds moral value. As such localization is deemed necessary to attribute value, the conclusion drawn is that technology does not inherently possess moral values (Miller, 2021).

The dual-use argument supports the neutrality thesis from another angle. It highlights the idea that technology can be used for both good and harmful purposes. For example, when a hospital is alerted, a fast car is dispatched to assist a patient, positively impacting their well-being. However, the same type of vehicle can be used for harmful purposes such as human trafficking or terrorist attacks. Similarly, drones and cameras may be used agriculturally to identify diseased crops, but they can also serve in espionage. Another case is mobile apps, which can be instrumental in safeguarding public health but may also be employed for mass surveillance. A stark example occurred in 2019, when it was disclosed that an app (the Integrated Joint Operations Platform) was being used in China's Xinjiang province to monitor and discriminate against millions of Uyghurs and other Turkic minorities.

These examples are about technologies that can be used in two ways: both for good and for bad. Applied to the topic at hand: they are about technologies that can be used both in favor of and against a particular moral value. But should we not go a step further? For what applies to drones and apps, does this not apply to *all* technologies? Even weapons—artifacts designed to cause harm—can be used for both a good purpose (defending a group of people) and a bad purpose (terrorizing a group of people). Indeed, it is difficult to think of a technology that one cannot use for opposing purposes, and for some this is an argument in favor of the neutrality thesis. After all, if an artifact can be used both for and against a value, should we not conclude that it is, in itself, value neutral, regardless of its good and bad effects?

At first glance, this conclusion seems to be justified. Consider the scenario of manipulating national elections: you need a comprehensive understanding of the electorate, so you develop an app distributed via social media to collect vast amounts of data from users unknowingly. This example refers to the Cambridge Analytica scandal, where there was a clear breach of privacy (Kaiser, 2019). If one were to argue that this technology is not value neutral, for instance, because one of the values attached to it is protection of privacy, it would seem contradictory to use the technology in a way that violates privacy. Similarly, it would be contradictory to use technology aimed at promoting equality while also claiming it was designed with racial segregation in mind. Given that virtually all technology can be used both to uphold and violate moral values, some resolve this contradiction by asserting that technology is value neutral. How persuasive do you find this argument?

The Value Ladenness of Technology

From the Stoics to today's technology gurus, some suggest that technology is neutral. In the pages that follow we want to argue against that dogma. Technology is not necessarily value neutral; we conclude that some technology is value laden. But before we get to that, we first provide examples of technologies that clearly carry values. Although these values might be morally neutral, they still raise questions about the assertion that technology lacks any moral value.

Survival and Living Well

Earlier we pointed out that there are two kinds of values: moral and morally neutral values. An example of a morally neutral value is efficiency. It is often attached to technology (but not necessarily). The materials are then chosen and composed in such a way that the use of the technology does not involve unnecessary losses. Of course, this is not an argument against the neutrality thesis. After all, that thesis only states that technology is not loaded with moral values. Nor is it an immediate reason to doubt the neutrality thesis. Indeed, efficiency is very different from moral values such as justice or privacy.

Besides efficiency, there are other morally neutral values. Take the value of truth, for instance. It can have a moral character (think about lying), but it can also be morally neutral. This is the case in the context of science, in the sense that researchers strive for theories that are true. Scientists strive to formulate claims that correspond to reality: for example, they want to find out the genesis of Chinese vases or the causes of gender inequality. Often, technologies play an important role in this context. Think of telescopes, atomic clocks, microscopes, particle accelerators, or computers. The first sort of camera also belongs in that list, as it was originally created to gain insight into the walking behavior of horses. All of these technologies exist with truth in mind, and so they are loaded with value, albeit with a morally neutral value. But if technologies can be loaded with this type of value, then why not a moral one?

Some may still not see this as a reason to doubt the neutrality thesis, arguing that truth is fundamentally different from values such as fairness or privacy. However, is the distinction between a moral and a morally neutral value always significant? Consider medical technologies such as pacemakers, infusion pumps, stethoscopes, defibrillators, MRI scanners, or the smart contact lenses developed by Google and Novartis for diabetics. These lenses have a chip that measures blood sugar levels through tears and changes color to indicate if sugar intake or insulin is needed. Clearly, these technologies are designed with a specific purpose: to prevent, manage, and treat illnesses, inherently carrying medical value aimed at preserving health.

If technologies can embody such medical values, why could they not also hold moral values? Yes, health is a medical value concerned with survival, whereas moral values relate to quality of life. But are these differences truly substantial, or are they simply distinctions without a deep difference? Could it be that proponents of the neutrality thesis are mistakenly treating ethics as an outlier, ignoring how deeply values can be integrated into technology?

The Moral Duty of the Designer

We have arrived at the core of our story. We appeal to three arguments to refute the neutrality thesis: the precautionary argument, the side-effects argument, and the enhancement argument. What exactly do these arguments entail?

Designers and engineers have no duty to increase the well-being of users or stakeholders. If they contribute to welfare, we praise them, but if they do not, they are not punished. They do, however, have to try to get the best possible view of the potential undesirable side effects of the technology they are creating, and to make sure as much as possible that those effects do not happen. Accidents happen and some undesirable consequences cannot be foreseen. But it is the engineer's duty to take precautions if it appears that there is a real risk of harm in using the technology.

Designing technology involves more than technical expertise; it also encompasses ethical considerations. Take, for example, the design of medical technology. If you know the equipment will be used both for therapeutic purposes and for collecting patient data for scientific research, it becomes essential to design the technology to anonymize these data. This means ensuring that personal identifiers such as names, dates of birth, and places of birth can be removed, effectively separating patient data relevant to science from personal identity.

Similarly, consider the development of algorithms used in recruitment or lending. If there is a risk of these algorithms discriminating based on ethnicity, gender, age, disabilities, or appearance, it is crucial for programmers to develop these algorithms in such a way that prevents moral problems such as racism or sexism. These examples underscore the importance of integrating ethical principles into technology design to address and mitigate potential ethical issues.

If these risks exist and the designer then effectively builds mechanisms into the technology so as to prevent them, then you can raise this sort of technology as an argument against the neutrality thesis. Certainly, the value ladenness is purely negative in this context. After all, the technology is implicated in a moral value only to the extent that it is designed so as *not* to threaten that value. Yet negative involvement is a reason to reject the claim that technology is value free. If a technology is designed so that it must avoid threatening a moral value, then the technology gives reason to believe that the use of the technology will be aligned with the value; being designed for a value implies that there is a high probability that the use of the technology will be in line with a value, meaning that the technology is laden with a value. This is the precautionary argument against the neutrality thesis.

British Telecom's Phone

Preventing something from having an undesirable effect on a moral value is not the same as ensuring more of that value. Preventing injustice differs from striving for more justice. The first intervention has a negative character, the second a positive one.

It goes without saying that some technologies have such a positive character. Numerous artifacts are used for the purpose of greater justice or autonomy. Importantly, even this positive link between a technology and a moral value is not merely established when the technology is used. That bond is often already ingrained in the technology itself, just like the negative bond we discussed in the previous section. This is what the side-effects argument is all about, the second argument against the narrow interpretation of the neutrality thesis. Take inclusiveness. This is a moral value that refers to the inclusion of people in a society or organization regardless of their age, sexual orientation, skin color, or other factors. It is clear that the use of certain technologies has undesirable effects on inclusiveness. This is not only evidenced by guns; ATMs higher than a wheelchair or buildings without elevators also exclude people. Yet you cannot deny that certain technologies are also linked in a desirable sense to the moral value of inclusiveness. We think, for example, of the company British Telecom's telephone from the 1980s. It was designed with large keys with numbers on them. The aim was to enable the elderly and people with limited motor and visual abilities to communicate at a distance. Some computer games also fit within this framework. Take *Mass Effect* and *Dragon Age* from the Canadian company BioWare. The version of *Mass Effect* released in 2012 included a male character whose sexual preference is for other men. It is not that gay characters are new to the gaming world, as games have been made in the past with women who fell for other women. The new thing about *Mass Effect* was that it introduced a male homosexual character. In *Dragon Effect*, in turn, there are characters with bisexual preferences and attention is paid to transsexuality. The goal is inclusivity, to create space and respect for all identities both inside and outside the gaming world, and not just according to the heterosexual norm (van de Poel & Kroes, 2014).

Telephone and computer games differ from technologies that are value laden in a negative sense, such as medical technology that disconnects patient data from individuals. The latter is equipped with a mechanism to prevent a certain morally desirable state from being threatened. In contrast, the telephone and computer games have a positive character, in the sense that they are aimed at greater inclusiveness. Yet there is also a similarity between medical technology and the telephone and computer games. In each case, the effects are intended by the designers, meaning that the construction of the technology is such that one can expect use of the technology to be aligned with a moral value. This orientation of the design toward a desirable state implies that there is a moral value attached to the technology, independent of its use. This positive involvement is the second argument against the claim that technology is not value laden.

Improvement Through Technology

The second argument is that some technologies are designed for greater fairness and privacy. That effect is intended; in other words, it is inherent in the design. Yet the link between technology and moral values does not cut deep here. If you replaced the keys and digits of British Telecom's phone with smaller ones it would eliminate a specific type of phone, but a phone with smaller keys and digits is still a phone. It is not suddenly about a different type of technology. The effect, inclusiveness, may be intended, but it is not essential to the description of the technology. This also explains the name given to this argument: the side-effects argument.

Some technologies are made primarily for the purpose of increasing a moral value. That is what the enhancement argument is about, the third reason for rejecting the neutrality thesis. An example of such technology is Tor, also known as The Onion Router, which was originally developed with support from the Defense Advanced Research Projects Agency and the US Navy. What is Tor?

When you browse the web 'normally,' your location and IP address can, in principle, be tracked without any problem. If you visit a certain website, it is usually not very difficult to trace the physical location of your computer. Software such as Tor allows you to surf completely anonymously. You do not have to do much more than download and install the Tor browser to do that, just as you would with Chrome and Firefox. As a result, you are not linked to one server, as is normally the case, but to a network of servers. If you enter a web address, you will be led along this long chain of servers. The point is that each server in that long chain only keeps routing information from the previous and the next server in the chain; a server cannot discover the data from other servers in the chain. The result is that the routing information arrives at the destination (the website) 'bare,' as it were. That is, you cannot be traced from the website you visit; it is then impossible to find out from which IP address the website was visited. This is also why Tor has an onion as its logo. The software works with several layers of protection. If the layers are removed, you end up seeing nothing at all (Warnier et al., 2015).

Anonymous communication is very interesting for military organizations and explains why the US military supported the development of the software. Meanwhile, Tor is used by journalists, political activists, or whistleblowers such as Edward Snowden. One can criticize and take action without the risk of detection, prosecution, or arrest. Does that mean that Tor is without danger? By no means. The software is in high demand by hackers and cybercriminals. And that has exactly to do with the point we want to make. Some technology is designed with mechanisms to prevent the violation of privacy (negative value ladenness) or is designed with a moral value in mind, but without that value being essential to the characterization of the object in question. Other technology, however, such as Tor, is designed primarily to ensure that there is more of a certain moral value, such as more privacy (positive value ladenness). It is precisely this surplus of anonymity that is so attractive to hackers and criminals. And it is also for this reason that one can use Tor against those who defend the neutrality thesis.

Moses' Mistake

Based on the past paragraphs, one might get the impression that all value ladenness is desirable. And *usually* it is, but it is not always the case. We pointed this out earlier: some moral values may be desirable for some people but undesirable for the majority. Think of eugenics from the nineteenth and twentieth centuries, and mainly the Nazi version of it. That was a form of research aimed at strengthening the Aryan race and eradicating races that were supposed to be inferior. Of course, that goal is reprehensible, but racial segregation was something that was seen as desirable in Nazi Germany. In other words, racial segregation was a moral value at the time. Conclusion? The confluence of science and moral value can have dramatic consequences and thus can be completely undesirable.

What is true of science is also true of technology. Technologies can be value laden in a morally reprehensible sense, regardless of the use phase. By far the best-known example comes from political scientist Langdon Winner in his famous text 'Do Artifacts Have Politics?' (Winner, 1980). We know that Winner's example is almost worn out by now, and we realize that possibly not all the information in his text is equally reliable, but we can use his case as an interesting thought experiment. Therefore, let us take a moment to consider the issue to which Winner wanted to draw attention.

In his text, Winner discusses a few projects by Robert Moses, the American urban planner who had a great influence on the urban development of New York and its surrounding region between the 1930s and 1970s. More specifically, he zooms in on the famous Parkway bridges over the highway heading toward Long Island. The reason why that bridge construction attracted Winner's interest is that the overpasses were built particularly low over the highway. Although that could be the result of incompetence on the part of the engineers, this was not the case here. The bridges were intentionally constructed that way by Moses. Why? The bulk of the African American community in New York did not have a car to drive to the beach on Long Island. They therefore generally took the bus to get around. However, because the bridges were built so low, it was impossible for those buses to pass under them and transport New Yorkers to the beach. As objectionable as that is, it was exactly what Moses had in mind. The ultimate goal of his design was a beach with only white people on it. In other words, the purpose was racial segregation.

This case is a variant of the side-effects argument, the second reason for rejecting the neutrality thesis. In the case of British Telecom's phone and BioWare's computer games, we are talking about desirable moral values, and now we are talking about undesirable ones. The similarity is that the value is part of the technology, and the value engagement does not cut particularly deep: it is not essential to understanding the technology. A bridge that does not discriminate is still a bridge, just as a phone without big keys or numbers is still a phone.

Repulsive Things

In addition, are there technologies where being laden with an undesirable value is indeed crucial to defining the technology? Does the enhancement argument also have a variant that is seen as good by some but morally repugnant by the majority?

No one would deny that the Nazi furnaces were artifacts, but some would suggest that those artifacts are themselves morally neutral. One cannot make moral judgments about the technology (and the designer), it is sometimes believed, but only about those who use the ovens. Yet, there is good reason to contradict this. After all, the design of these ovens differs greatly from the regular ovens used to cremate bodies. They do not have aesthetic ornaments, they have more silencers, there is no mechanism to distinguish between the ashes of different corpses, and the ovens have more corridors to speed up the process (Miller, 2021). In other words, the design makes it clear that the ovens were designed for a reprehensible purpose: mass destruction. Burning many corpses is not something you can also do with the ovens, separate from the main purpose. Rather, it is their main purpose, as is evident when you look more closely at the ovens. This not only makes it absurd to claim that you can only make moral judgments about the user and not the designer, it also shows again that some technologies are value laden through and through. Software such as Tor is desirable (for many people); the ovens are atrocious, nauseating. Nevertheless, they correlate in the sense that the design of the technology makes it very likely that the use of it will be aligned with the disvalue. This makes the Nazi ovens (just like atomic bombs, by the way) a variant of the enhancement argument.

This concludes the core of our argument. To summarize: those who defend the neutrality thesis argue that technology cannot be laden with moral values. We have given three arguments against this thesis. The similarity between them is that they each start from the designer's point of view of realizing a moral value. This orientation translates into the design of an artifact, so that the artifact contains a value and so that you can no longer claim that technology is separate from a moral value. There is not necessarily a gap between technology and moral value, we conclude; they can be intertwined. The difference between the arguments has to do with the type of value ladenness (negative or positive) and the degree of value ladenness (weak or strong). The first argument recalls that some technologies are laden with a moral value in negative and weak ways. There are mechanisms to prevent the technology from having undesirable effects—that is the negative aspect—whereas those mechanisms are not crucial to defining the technology—that is the weak aspect. The other two arguments refer to the fact that the goal of some technologies is to contribute positively to a more just world, for instance. Sometimes that effect is not essential to the characterization of the technology (second argument), and sometimes it is (third argument).

Some Critical Notes

It seems right to put the brakes on at least two occurrences now. When it turns out that technology is not necessarily morally neutral, there is a risk that we will err on the side of caution and proclaim the opposite: all technology is value laden, no technology is value neutral. Another risk is that designers believe that equipping technology with a moral value cannot create additional problems. After all, why would a moral value create new problems? In the following pages we explain how to protect against both risks.

The Neutrality of a Drill

We have already pointed out that designers have a duty to consider what undesirable consequences might result from the use of their design. If such risks are found, then mechanisms must be built into the design to prevent such effects. Technology is then value laden in a negative sense, as we now know. However, is technology *necessarily* value laden in this sense?

The answer to this question is negative for two reasons. First, technology can be made without considering its potential undesirable consequences. Taking precautions is not sufficient for making technology, but nor is it necessary. This implies at least the possibility that a designer, although obviously morally highly reprehensible, may neglect his or her duties during the design phase and not consider the potential harmful effects. In that case, one cannot claim that the artifact is laden with a value in a negative sense. Second, it is also possible that the designer does consider the potential undesirable effects during the design phase, but that she or he believes that the use of the technology will not be accompanied by harmful effects on values such as privacy or fairness. In this case, the design is not changed according to possible undesirable consequences, and the artifact itself does not give reason to believe that the use will be in line with a moral value. Many 'primitive' technologies are examples of this: scissors, hammers, knives, forks, brushes, among other basic things.

What about the other type of value ladenness? Are all technologies loaded with value in a positive sense? Are they all designed with a view to realizing a moral value?

We have already seen that all technologies have instrumental value. This is the case because they are designed for a purpose. Furthermore, it is also true that many technologies are value laden in a positive sense. Consider the examples we mentioned earlier: Tor, computer games, British Telecom's phone. And yet, not all technologies are value laden in a positive sense. Take a drill, for instance. A drill is an artifact designed to turn a screw in a piece of wood or stone, or to make an opening in a wall. But making an opening in the wall or turning a screw in a piece of wood are not values, let alone *moral* values. Of course, safety mechanisms may be built in, making the design refer to the value of safety. But that value ladenness is not necessary for a technology to be a drill. For a technology to be a drill, it is sufficient that it be designed to turn a screw. And as that purpose is not a moral value, one can say that such an artifact is morally neutral.

Some answers are as follows. Even if the direct purpose of a drill is not itself a value, that purpose is always connected to a value. Admittedly, if that is correct, then it becomes very difficult to deny that all technology is value laden in a positive sense, even if the link between the technology and value is indirect in that case. But is it true that the purpose of any technology is indirectly linked to a value?

Take the drill again. You can use that artifact to hang a coat rack on the wall, for example. In that case, there is indeed a relationship between the drill and a value such as comfort or order. But you can also use that machine for other things, for example, to hang a cuckoo clock in the living room. Does the drill then contribute to the realization of a value? Tom bought that clock because he likes the fact that every hour it imitates the sound of a cuckoo. He also finds it very funny that his friends are startled several times when they visit him. But does that also lead to something important, something that carries so much weight that it must be realized? Does that also increase his happiness or well-being? Anyone who answers that in the affirmative is confusing values with things that are just fun. No one is saying something has to be fun to be a value, but one thing is clear: it is not sufficient that something needs to be fun for it to be a value.

So thinking about the relationship between technology and value requires some nuance. Not everything is value neutral, as we already showed. But now it also appears that not every technology is value laden; not all machines are moral machines. There is no necessary relationship between technology and moral value, just as technology and neutrality are not inextricably linked. Some technology is value laden, and some is morally neutral.

And Justice for All

Earlier, we mentioned that we should apply the brakes twice. The first pause concerned the neutrality of some technologies. The second pause addresses the following: it is almost a given that designing technologies is a complex process fraught with various challenges—technical and logistical issues, for example. But do these encapsulate all potential design problems? Not quite: moral values often introduce additional complications. We discuss two such moral dilemmas.

One is the so-called 'confusion of tongues problem' (Brey, 2015). Consider justice and autonomy—values widely regarded as essential, and most would agree it is our duty to strive for them. However, does this consensus hold when we delve into what these values specifically mean to different people? The answer is no. Take the concept of fairness, for instance, which is understood in our society in at least two distinct ways: fairness as statistical equality and fairness as equal opportunity.

Statistical equality can be illustrated by the distribution of event tickets between two groups, say men and women. According to this view, fairness does not require an equal number in absolute terms. If there are 50 men and 10 women, it would be fair, under this interpretation, for more men than women to receive tickets, and unfair if men and women were to receive an equal number of tickets.

The second interpretation, equal opportunity, emphasizes that individuals should begin from as equal a starting position as possible, such as when starting university studies. Later differences in wages, under this view, are less concerning than unequal opportunities that stem from different socio-economic backgrounds at a young age. In this case, fairness might involve providing a scholarship to one student and not another to help to equalize or at least reduce these initial disparities.

Increasingly, even within tech circles, there is a growing focus on fairness, which is undeniably positive. However, this raises an important question: what do we actually mean when we say that a technology, such as an AI system, is fair or unfair? This concern extends beyond fairness to include other moral values such as privacy and sustainability. Why, then, is this conceptual clarity crucial in the realm of technology?

Suppose the government allocates funds for developing AI systems to award scholarships and hires a company with explicit instructions to ensure that the AI system is fair. Given that fairness can be interpreted in multiple ways and involves two stakeholders (the government and the company), there is a risk that different notions of fairness might be applied unknowingly. If, as posited, each party holds a different understanding of fairness and assumes a shared perspective, this could lead to a confusion of tongues, where discussions miss alignment without the parties realizing it.

This misunderstanding could have serious implications. When the ministry reviews the technology just before its market launch, they may find that the system does not align with their expectations owing to differing definitions of fairness. At best, this might cause minor friction; at worst, it could necessitate significant changes to the system or even a complete redesign, wasting both time and money on the assumption of a non-existent consensus about fairness. The key takeaway is that although the development of value-driven technology benefits from scientific knowledge, this alone is not enough. It is crucial to also recognize that moral values can vary in interpretation, and clear communication of each party's understanding of these values is essential.

Integrating moral values into technology can also lead to the so-called 'collision problem,' where values may conflict. Consider a bank that uses an AI system for loan approvals, analyzing two groups: Black women and white men. The AI, with perfect accuracy, predicts that 15% of Black women and 30% of white men will repay loans.

Despite its accuracy, the bank is concerned about potential discrimination and wants to ensure fairness, defined as equal loan distribution between the two groups. This decision impacts the accuracy of the AI: if the bank adjusts the AI to grant loans equally at 30% to each group, it creates false positives among Black women. Conversely, setting the rate at 15% for both results in false negatives among white men, who could repay their loans (Kearns & Roth, 2020).

This scenario illustrates the collision problem, showing how the moral value of fairness can negatively affect the morally neutral value of accuracy. The more the AI strives for equal loan distribution, the less accurate it becomes. The dilemma highlights that although implementing ethics in technology is crucial, it also necessitates understanding that different moral values can mean different things and may conflict. Therefore, it is important not only to recognize these potential clashes but also to consider carefully whether the pursuit of one value, such as fairness, justifies compromising another, such as accuracy. This involves a deliberate trade-off, where the values must be weighed against each other to determine priorities.

The Importance of Neutrality

Now that our critique is complete, we return to the defense of the neutrality thesis, the deeply held belief that technology has no moral values attached to it. Thus, the two arguments we have given—the visibility argument and the dual-use argument—fall short, in our view. But how exactly are they deficient? Why are they unable to do what they are supposed to do, which is to support the neutrality thesis? That is what we are going to look for first. Then we will ask ourselves whether defending the neutrality thesis has anything to do with anything other than arguments. Are there interests and deep-rooted, ill-considered ideas lurking behind this thesis?

Seeing Values

To begin with, we recall how the first argument for the neutrality thesis, the visibility argument, breaks down into two parts. The first part holds that technology can only be value laden if you can identify a moral value in its design. The second part holds that it is impossible to designate a place where you can actually see the value. For example, the aforementioned Pitt, a staunch defender of the moral neutrality of technology, writes the following when discussing the alleged value ladenness of Moses' bridges: "Where are the values? I see bricks and stones and the footpath, and so on. But where are the values—do they have colors? How much do they weigh? How big are they or how skinny?" (Pitt, 2014). Why is the visibility argument problematic?

The problem with the argument is not of a logical nature. If the two parts of the argument are correct, then one must infer that technology is morally neutral. The question, however, is whether both parts are correct. In any case, Pitt is right about the second part. When you look at technological artifacts, you see buttons, wires, keys, ports, glass, iron, handles, and other things. But do you see values in the object? Sure, you can see that the material was chosen and put together to realize a certain value. But you cannot pinpoint values such as sustainability or privacy within technology itself. This is actually self-evident. Technologies are things, usually of a material nature. Moral values, on the other hand, are not things. They are often about a person's life: think about autonomy, or about the relationship between persons, for example, in the case of fairness.

So, the problem has to do with the first part of the argument, which is that you have to be able to pinpoint a value in an object. If that is impossible, then that object cannot be value laden. Clearly, this requirement is based on a particular interpretation of 'value laden.' Earlier in this chapter we pointed out that this term can be interpreted in two ways. If you interpret it figuratively, it means that the material or design is such that one may expect the use of technology to be in line with moral values. In that case it is not necessary that you actually perceive a value in the technology to call it value laden. A literal interpretation of 'value laden' means that the moral value is part of the design, just as cables or pipes are. Evidently, the visibility argument is based on this interpretation. Indeed, the requirement that you must be able to see a moral value in a thing before you can call it value laden only makes sense if 'value ladenness' means that a value actually belongs to the design.

So, the question is not so much what the first part of the argument is based on, but rather *why* the literal interpretation is chosen. One would expect to have good reasons for doing so. Indeed, the consequence of this choice is that technology cannot be seen as anything other than morally neutral, that it cannot be value laden. We have already pointed this out several times: technologies and moral values are two completely different things; thus, a moral value cannot possibly be in a technology. The problem now is that those who defend the neutrality thesis on the basis of the visibility argument do not justify their choice of the strong interpretation. Moreover, there is no good reason, in our view, why one should not choose the figurative interpretation and instead choose the literal interpretation. We are not claiming that there is no good reason. We only point out that no reason is given and that we do not see such a reason. So the problem with the visibility argument, in our opinion, is that it starts from an unfounded decision.

One can also criticize the argument by pointing out the following. We say of many things that they are value laden. Even those who claim that technology is value neutral think so. All or most people think that scientific research aimed at reducing discrimination, for example, is value laden. Yet you cannot actually pinpoint the value of equality in the practice of science. Equality plays out between the people being researched, but not in books, studies, or interviews. So in order to be able to say that research is value laden, it is apparently not necessary for many, including those who defend the neutrality thesis, to be able to point to a moral value. So why would this not be necessary when it comes to science and other things, but necessary when it comes to technology? Is there a relevant difference between the two that makes the stringent requirement apply to technology and not to science? Those who invoke the visibility argument offer no further explanation for this, and in our view, this is because there is no relevant feature. It is possible that such a characteristic exists—we want to underline that here too—but it has never been invoked and we do not see what characteristic it could be. So again, the problem is this: the argument is founded on an unsupported choice.

We are not claiming that the visibility argument is based on nothing. We suspect that it is theoretically unsupported, but it could be that there are nontheoretical motives at play, such as vested interests, for instance. In other words, it is possible for one to choose the literal form of value ladenness and thus the inevitable consequence of seeing technology as value free, not only because one has arguments for it but because one wants technology to be seen as technology, for whatever reason. Here, at least, the visibility argument is based on what we believe are unfounded decisions. More on this later.

Designing and Using Technology

The second argument—the dual-use argument—is based on the observation, as we explained earlier, that you can use (almost) all technologies in both good and bad ways. That double use, so the reasoning goes, necessitates the conclusion that technology is value neutral. Indeed, can you claim that an artifact is loaded with a value such as privacy, while also being used to violate someone's privacy, and vice versa? No, according to some. And because all or almost all of the technology that was developed for a value can also be used against that value, one must conclude that technology must be value neutral.

The dual-use argument is founded on a well-defined belief. Underlying the argument is what is known as a deterministic assumption. That is, it is assumed that the design of the technology fixes its use. It is assumed that loading a design with a morally good value implies that you cannot use the object to achieve its opposite. It is impossible, it is believed, for an artifact to be laden with a certain moral value while being used in such a way that contradicts that value. The reason why the dual-use argument now does not do what it is supposed to do, which is to support the neutrality thesis, is that the deterministic assumption is wrong.

Let us make a comparison to clarify this. Health is a value, not a moral one but a medical one. You can use resources to benefit your health, but you can also harm your health, for example, by smoking. With some artifacts you can do both. You can use a hypodermic needle to prevent or cure a disease, but with that same instrument you can also inject substances that do not benefit your health. Yet this does not necessarily mean that the hypodermic needle is value neutral. It is possible that it was made with the medical value of health in mind—the hypodermic needle is then value laden—while being used not only for good but also for bad. So, when it comes to a medical value, the deterministic assumption is incorrect: the technology can be value laden, but it can also be used in such a way that is radically against that particular value.

Why could this not also be true with moral values? Of course, medical values have their differences, but are they such that we cannot reason in the same way when it comes to privacy, fairness, or sustainability? Take a weapon, something that can be used in a dual sense. On the one hand, you can use it to liberate people. In that case, it has a desirable effect on the moral value of autonomy. On the other hand, one can also use it against that value, for example, by terrorizing people with it. But does that imply that the object is value neutral? No, the weapon may have been designed with the intention of liberating people. It is then clearly laden with a moral value, whereas that loading does not prevent it from being used in a morally reprehensible sense, for example, to oppress people. Conversely, you can make a weapon with a terrorist purpose in mind that is nevertheless used afterward in a morally correct sense. Therefore, that proper use does not make it value free. It is and remains laden with a value, even though that value now has a morally reprehensible character. Conclusion? The deterministic assumption holds water neither on the medical level nor on the moral level, which is also why the dual-use argument does not lead to the conclusion that technology is morally neutral.

It's the Economy, Stupid!

We now have a good view of the problems with the argument for the neutrality thesis. Nevertheless, we do not want to give the impression that whoever reads our exposition will no longer defend this thesis. No matter how extensively we have dwelled on that thesis, it is possible that arguing it may have little or no effect on those who say that technology is morally neutral. Why is this the case? Why is it that philosophy hits a wall here?

When people are guided *solely* by arguments in a debate, they rally behind the assertion with the strongest arguments. This means that they are receptive to counterarguments, through which they may revise their opinion. But assertions can also be defended on the basis of things other than arguments. We are thinking, first and foremost, of vested interests. Some people subscribe to a view, not only because they may have good arguments for it, but also because they have an interest in it, because something is at stake that has nothing to do with theory or insight. The possible consequence is that people are not open to arguments that undermine their own beliefs. When an interest is at stake, people may cling to a proposition to such an extent that they will continue to support it, even if it turns out that it is not supported by the best arguments. Could it not be that this is also the case when it comes to technology, or more so, precisely because it is about technology? Is it not the case that some say that technology is value free because they have something to gain from it? Some may defend the neutrality thesis purely on the basis of theoretical arguments, but the possibility that we want to put forward here is that others are also guided by nontheoretical motives, or indeed, that for a number of people such motives are the only factors that come into play. The interests that we have in mind here are moral and economic. Let us clarify this.

Numerous technologies have an undesirable effect on moral values: AI systems discriminate, cars emit greenhouse gasses, and apps violate privacy. When those effects were unforeseeable, we do not pass negative judgment on the designer, manufacturer, or stakeholder. But they are not necessarily left out of the equation. When the bad effects were foreseen, even if they were not intended, this is a reason for a negative evaluation. Furthermore, you can also condemn those involved if it turns out that the undesirable effects of the technology were not only foreseen but also intended. This is the case with some of the examples mentioned above. Although we are more likely to praise the creators of the computer games just now, we condemn the designers of the Parkway bridges, just as we pass negative judgment not only on the users but also on the designers of the Nazi ovens. The reason is the same each time: the objectionable effect is embedded in the artifact, and thus does not result solely from the user's bad intentions. The effect is also directly linked to the developer, and that is a reason to condemn them.

Looking at the neutrality thesis from this point of view at least sheds new light on the defense of that thesis. It shows that the defense of the neutrality of technology may also stem from a desire to wash one's hands of it. We are not referring here to those who equip technology with morally good values such as fairness or sustainability. We are referring to the designer, manufacturer, or stakeholder who pursues morally wrong values such as racial segregation and discrimination and who develops technology in service of those goals. Only the latter does not want to be the subject of disapproval itself; it is the user who must bear all the moral blame. The individual pursues a moral evil, but at the same time desires to wash his or her hands of it. See here the desire for clean hands that can be hidden behind the neutrality thesis. I am not to blame, seems to be the message behind this thesis. I did design and create the technology from which moral evil arose, but that evil is not embedded in the technology. What I designed is morally neutral, and so you have no reason to condemn me morally. You should condemn the user, because moral evil flows from him or her, not from the design. Although obviously not justified, that is the reasoning of the designer or manufacturer that might be behind the defense of the neutrality thesis. It is also the reasoning that makes the link between this first chapter and the next, in which we spend quite a bit of time examining the question of moral responsibility.

There is, in addition, another interest that might lie beneath the neutrality thesis, an interest that is economic in nature. To introduce that, we refer to the following. In a liberal state, governments must be neutral. Not only may they not express a preference for a religion or sexual orientation; if there is no good reason to do so, it is also not permitted to favor or disadvantage certain technologies. But there is a limit to that principle. If it turns out that the use is accompanied by severe unforeseen negative consequences, then one has a reason or even a duty to take the technology off the market. The government can intervene even more quickly, for example if it is already clear during the design phase that very undesirable side effects will occur. However, there may be another reason to intervene. In addition to being merely foreseen, the bad effects can also be intended by the creator, designer, or stakeholder. This is so if the design is laden with an objectionable value such as inequality. In that case, the bad effect is embedded in the technology and the artifact was designed with that bad purpose in mind. Moses' bridges are again illuminating in this context. The effect of using the bridges is that there are only white people on the beach. That effect is not a side effect, but the purpose of the bridges. Moses made the bridges so that the beach would be populated with only white people. That is enough to make a negative moral judgment on those involved, but it is also a reason for the government to stop the development.

Looked at this way, it would not be surprising if it turned out that the neutrality thesis is kept alive by industry, and that it is particularly the designers and producers of morally highly questionable technologies who continue to defend it. On the one hand, they know very well that their technology is laden with wrong values and that this is sufficient for a ban, but on the other hand, they have an interest, an economic interest, in ensuring that the ban does not occur. A brake on production means no income and profit. Of course, it is not excluded that industrialists will invoke valid arguments, but it is also not excluded that interests are hidden behind those arguments. It is legitimate to prohibit the use of technologies that produce undesirable effects, but because production must continue and there is a risk that it will be curbed, we will subscribe to the idea that technology is morally neutral. At least, that would be the line of thinking that we suspect some entrepreneurs share.

Could it be that such vested economic interests are at the heart of 'Guns don't kill people, people kill people!,' the slogan with which we opened this chapter? It is usually explained as follows. There is a relationship between guns and things of which we morally disapprove. However, that relationship only exists because there are people who use guns in evil ways. The evil is not contained in the weapon itself. Weapons are neutral; there are no morally wrong values embedded in that technology. As a result, you have to ban or at least regulate the use of weapons; you have to leave the development solely to the market. We will not go into the details of whether the latter makes sense. What is certain is that the explanation behind 'Guns don't kill people, people kill people! ' is rarely argued. Furthermore, in this context it is also relevant to recall that the slogan was first printed in *The Manufacturer*, a magazine for factory owners.

To avoid mistakes, we do want to underline the following. If it is true that the neutrality thesis is not—or not exclusively—based on theoretical arguments, but (also) on all-too-human motives, this is no reason to reject the neutrality thesis. A theory does not stand or fall depending on the kind of motives from which it springs. Statements are not false because they arise from an agenda, radically left-wing or right-wing or otherwise, because there is money to be gained or lost, or because one is jealous or envious. A theory stands or falls because there is nothing or something wrong with the theory itself respectively. Yes, a theory can be wrong, and it may be that one defends the theory out of self-interest, but the theory is not wrong *because* it stems from self-interest. Keeping that in mind, it becomes clear that we referred to interests just now, not to challenge the neutrality thesis, but to explain it. More specifically, we want to make clear why such a claim has been around for so long and why it is so popular with computer scientists, entrepreneurs, manufacturers, engineers, and others intimately involved in the design and creation of technology and AI. Our proposal is this: some people defend the neutrality thesis, not so much because they believe it, but because they have an interest in doing so, because they have something to gain or lose, because something is at stake for them.

Machines Are Not Rocks

A theory of technology may be clouded by at least two things: bad arguments and vested interests, or a combination of both. Finally, we would like to put forward a third possibility. The defense of the neutrality thesis may also result from looking at reality through a modern lens. For this we appeal to the thinking of philosopher Bruno Latour (2012).

It is well known that in the seventeenth century, René Descartes questioned whether we should trust our observations and mathematical knowledge. He doubted everything, with the result that he is sure of at least one thing, which is that he is doubting, *ergo*: I exist. After all, it is impossible to doubt but not exist at the same time. The entirety of Descartes' thinking after this consists in solving the so-called bridge problem. That is, he must bridge the gap between his own existence and the world around him, the outside world. For although you may be sure that you yourself exist, is there also a world outside of yourself? The outline of his question is thus: here, on this side, I stand, and I would like to believe that there, on the other side, is an outside world, but how can I find out with full certainty? This is the distance Descartes wanted to bridge.

According to Latour, this contrast between man and the world is typical of a modern way of thinking. It is not that philosophers have not previously asked themselves the question of whether the world exists, but the fact that Descartes formulated the question so explicitly in those terms is far from accidental. After all, from modernity onward—roughly from the seventeenth century—people began to think more and more in terms of opposites. Not only was man placed opposite the world, but nature versus culture, values versus facts, were also put in opposition. To be modern, says Latour, means that you place things opposite each other and draw sharp lines; it means that you think in terms of binary. In contrast, there is a premodern worldview, in which things are less clearly distinguished. Nature is not seen there as something opposed to man, but as an order in which man must occupy the place reserved for him. Take lightning and thunder. Although the moderns would understand these things in a physicalist sense, from a premodern point of view they are interpreted as signs, as signals of God's anger, for example.

Does that mean that the modern view is the right one? No, says Latour. Since modernity, people have indeed thought in terms of opposites, but reality is not so easily pigeonholed. The modern opposition between inside and outside, facts and values, nature and culture, is an illusion, a false narrative that we tell ourselves. *Nous n'avons jamais été modernes*, says the title of Latour's well-known book from 1991. Something is never purely culture or nature, inside or outside; everything is a hybrid, Latour believes, both culture and nature. Think of scientist Robert Boyle. To get a good look at nature, he had to control his vacuum pump, invoke witnesses, and seek support from political authorities.

We leave open here whether Latour's vision is correct, or whether everything is really a hybrid. Nevertheless, we would like to raise the following question. Could it not be that the neutrality thesis is a consequence of such a modern view? Does one mistakenly assume that there are no moral values in technologies because one is in the grip of the modern tendency to purge, to separate what is intertwined, to think in binary terms? In other words, is the claim that technology is value-neutral not itself biased? We advance this statement on the basis of the following.

Often the organic and the inorganic are kept apart. The former includes humans, plants, people, and other animals. Opposite them, at least as modern thinking posits, are nonliving entities such as iron and stones. We cite that distinction because the neutrality thesis fits into that scheme. First, the neutrality thesis also expresses an opposition between technology and moral value. Further, technology usually consists of dead substances (iron, steel), and there is a strong connection between moral values and the category of life (of human beings). After all, a certain condition is only a moral value for people, and such value is usually about the lives of individuals or interpersonal relationships. These issues provide the impetus to put forward the following possible explanation. The defense of the claim that technology is value free is based on a modern scheme of thought, an underlying framework that places technology alongside dead matter such as stones, and moral values alongside life and the organic.

We know by now that the neutrality thesis is incorrect; technology can be neutral, but it is not necessarily so. In terms of the modern paradigm: we know that the distinction between life, people, and values on the one hand and inorganic, dead matter on the other is not sufficient to capture the broad spectrum of technologies. The examples we have cited—think of Moses' bridges or British Telecom's telephone—are not reducible to dead matter. Those artifacts are hybrids in Latour's sense: they lie on the border between the inorganic and life, a border drawn too strongly in modern thinking. Nevertheless, the relations touched upon, between moral value and life, for example, are a reason to suggest that the neutrality thesis is an effect, not of arguments or vested interests, but of a modern view of the world. The claim that technology is separate from moral values may also result, we suggest, from the rigorous division of the world into life and dead matter.

The Point of a Philosophical View

What is the relevance of this? This is not only the question that is often asked of philosophers, it is also the question with which we want to conclude this chapter. We have zoomed in on the neutrality thesis and shown that it has many shortcomings, but is this line of reasoning also useful for practice? We believe that it is. Our argument has practical relevance in a number of respects: for the handling of technologies, for the evaluation of designers, and for users.

Praising Human Beings

As we noted in the Introduction, all technology is purposefully designed to perform a function. This means that a designer's work can be assessed based on whether the technology functions as intended. If a device is used correctly but fails to produce the expected outcome, the fault may be attributed to the designer; conversely, if it performs well, the designer is commended. This type of assessment is technical and does not involve ethics. However, designers can also be evaluated on legal grounds, which is distinct from technical considerations. For instance, a designer may be deemed negligent if they are unaware of relevant laws that they should know, or they may face penalties if their technology breaches regulations, such as those governing personal data. But do these technical and legal evaluations cover all the bases? Are there other ways in which we might assess AI developers and other tech professionals beyond these criteria?

These three types of judgments—technical, legal, and moral—can interconnect from a certain perspective. A designer might receive a positive moral evaluation that complements their technical and legal assessments. For example, one could create an AI system that not only functions well and adheres to all legal requirements but is also infused with morally commendable values. However, the situation can be more complex. Consider again the designer of the low Parkway bridges in New York, intentionally built to restrict access to a beach to predominantly white people. Technically, the work could be seen as successful because the bridges achieve what they were designed to do. However, morally, this design promotes racial segregation, which is undeniably reprehensible. Consequently, the moral judgment diverges sharply from the technical appraisal.

From this, we learn that moral evaluations of designers can lead to seemingly contradictory statements that, upon deeper analysis, cohere. It is feasible to describe the designer of the Parkway bridges as both competent and unethical. This apparent contradiction arises because the positive assessment refers to technical skills, whereas the negative pertains to ethical choices.

The Moral Relevance of Knowledge

In discussing the visibility argument, we already pointed this out: you cannot see moral values within technology. However, you may be able to notice value ladenness. One can see that the design is geared toward a moral value, or that certain materials have been chosen with a view to realizing a morally desirable situation. Yet it is not excluded that you do not see at all that a moral value is embedded in the technology, or that you hardly see it, for example, because the reference to a moral value is in the details of the design. Think again of the Parkway bridges. Those were undoubtedly laden with a morally objectionable value, but if you were only to look at those bridges, you cannot infer that value ladenness from the design itself.

So, if our argument is correct, it is possible to choose a technology without knowing that there is a moral value attached to that technology. In the first place, this is undesirable in itself, because everyone wants to make choices that are based on all relevant information. Yet this need not be accompanied by harmful consequences. After all, the moral value of which you are not aware may be good. What is worse is when technology is chosen that you are unaware is laden with a moral value, even though that value is objectionable. In that case, it is not certain that you will be held responsible for the consequences of that choice, but it is certain that using that technology as intended will be harmful. In addition, those consequences may be long lasting. After all, choices of technologies cannot always be overturned quickly, for example, because they are materially or spatially firmly entrenched, as in the case of the Parkway bridges. For this reason alone, it is relevant to know that technology is not always value free, as someone like Zuckerberg falsely claims. And that is also why it is highly advisable, prior to choosing or using the technology, to obtain as much information about it as possible. Only when this epistemic condition is met is the choice to roll out an app or implement an AI system truly morally sound.

Taking Care of Things

At the beginning of this chapter, we pointed out that we use the term 'value' in everyday life in two ways: something can be a value or something can have value. If an object has value in itself, then that is also a reason for a positive attitude toward that object, we pointed out. For example, one takes good care of a bicycle because the person likes the bicycle. But does that mean that every positive attitude follows from the fact that something has value? Can a positive attitude toward a technology also be based on something else?

One has to answer that question in the affirmative. If one knows that a technology is laden with a morally desirable value, then that is a reason to be positive about that technology. For example, if one is shown an AI system that is laden with fairness, that is a good reason to praise it or care about it. Of course, it may be that the person also cherishes or praises that system for reasons other than the fact that it is loaded with moral value—because it is aesthetically pleasing, for instance. In that case, the value ladenness is an additional reason for a positive attitude. But even if it turns out that a technology does not work or does not work well and that it has no other value, that does not mean, following our line of reasoning, that you can only have a negative or neutral attitude toward that technology. After all, that object may be laden with a moral value such as justice or equality, and that in itself may be enough to value it, care for it, or commend it.

Moral Machines

We will elaborate on that last point in closing this chapter. That a technology is laden with a morally desirable value is a reason to be positive about it, but can we not attach even stronger consequences to it? Do moral duties not also follow from our argument? We think that they do. This is why: when things are laden with a moral value, we have a duty to at least seriously consider their use. To explain this, we first briefly discuss the connection between technology and duties.

The fact that a bicycle, car, or watch is beautiful—read: that it has value in itself—is a reason to take care of it, but that fact in itself does not oblige us to do anything else. Yet we can also have obligations toward artifacts. Often, these obligations are based on ownership, on the fact that something belongs to someone. Because the bicycle belongs to someone, the person who uses it must take care of it; because the car belongs to Mark's neighbor, Mark is obliged to ask the neighbor whether or not he can use the car to drive to the store. In addition, Mark may have to take care of it because he promised the owner that he would do so. Duties can arise from ownership, but they can also arise from a promise. Both duties are clearly not embedded in the thing itself but are derivative of it. They derive from something other than the technology: from ownership or promise. In a sense, therefore, one could say that these two cases are about a duty, not so much to the object itself, but to a person: the owner of the thing and the one to whom the promise was made.

However, duties can also be based on things other than ownership and promise. Consider what we may and may not do with respect to human beings and animals who are not human beings. It may be one's duty to take care of a neighbor's cat because that person promised him or her to do so, but it goes beyond that promise. One also has a duty to the cat itself, independent of the promise. The same, of course, applies to our duties to other people (van de Poel & Kroes, 2014). One must take care of the neighbor's crying child, not only because one promised to keep an eye on him, but also and especially for the sake of the child himself. In these and other cases, duty is grounded in attributes, for example, the ability to feel pain. That ability carries with it the duty not to hurt a human or cat, or to try to take the pain away.

In Chap. 2, we go into this in more detail, but here it is enough to say that machines and robots do not currently possess the properties upon which our duties toward humans and animals are usually founded: the ability to feel pain and (self-)consciousness. Hence: we owe nothing to them. Our analysis from this chapter does not change this. From the fact that a technology is laden with a moral value, you cannot infer that you owe something to that technology itself. In other words, you cannot conclude that technology has rights based on our account in this chapter. Of course, value-laden technologies do have obligations attached to them, but those are obligations directed to the owner and not to the artifact itself—that was the point we were making just now. And yes, it cannot be ruled out that in the near or distant future we will find that we owe value-laden robots something, or that they have rights. But at that point it may not be because they are laden with moral value.

And yet we propose the following: value-laden technologies entail special duties. These are not, of course, duties with respect to that technology itself, but they are duties that derive from the properties of a technology, namely, the fact that a moral value has been pumped into the technology. Let us explain this further.

The moral psychology of humans is limited. We tend to care more about peers, and about people who are close to us in time and space. There is the so-called negativity bias: if we do not know someone or barely know them, we are more likely to notice their undesirable traits rather than their desirable ones. We think that we make moral judgments based on strong arguments, that we are nuanced moral reflectors, and that we readily admit mistakes. Recent research suggests that this is a story we like to tell ourselves, whereas in fact not much of it is true: good moral reasoning is less common than we think. Ethics usually flow from intuitions, emotional reactions, and ad hoc rationalizations (Savulescu & Maslen, 2015).

In the past and up until now, we have relied primarily on education and religion to improve our morals, to correct or eliminate moral deficits such as the negativity bias. Today, we could also turn to biomedical interventions. The beta blocker propranolol, for example, not only helps against high blood pressure, but would also appear to reduce racial prejudice (Savulescu & Maslen, 2015). Alternatively, we could rely on AI to transcend our moral limitations. Here are some of the options.

The number of hours we sleep affects our moral decisions—at least, that is what research on the judgment of sleep-deprived soldiers tells us. The same seems to be the case with hunger—we pointed it out earlier. Judges are more harsh when they are hungry. And environmental influences also affect ethics. When people are in difficult situations, they react with more hostility in high temperatures and in spaces with lots of people and noise. AI systems could be used in such contexts as feedback systems. They could collect physiological data and data about our environment, analyze it based on large datasets from the past in the light of the best possible moral functioning, and then send a signal to our cell phone the moment at which that functioning is in danger of being disrupted.

However, the use of AI for ethics need not be limited to mere monitoring. Such technology can also play a more active moral role. Take the following example. Many people intend to donate money to charities. Nevertheless, we know that people give less than they think. We could use AI systems to investigate what causes this gap and how the problem might be solved. Or consider fairness and bias. One study showed that knowledge of a person's sexual identity influences our judgment of that person. A woman's piano playing was better rated if the evaluators did not see the pianist. The judgment was significantly lower if they saw the woman playing. To eliminate such morally undesirable situations, we could employ AI systems that train people to make gender-neutral decisions.

Now suppose that such technologies are available, that we can make use of them. Then these are clearly examples of value-laden technologies. They carry the promise of more moral values, and of, say, less bias or more equality (Wallach & Collin, 2009). Do we not then at least have a duty to seriously consider the use of those technologies? We are not claiming that they should be allowed, let alone that they should be mandatory, but if those systems promise such morally desirable effects, is it not our duty to thoroughly consider and debate whether to implement them? Such innovations go against our moral intuitions. We usually think that we should leave moral training and improvement to people (parents and teachers). And of course, we know that a significant number of people have a rather pessimistic attitude when facing technology. A technology such as the microwave oven, for example, initially elicited very negative reactions; and research shows that many people are distrustful of and negative toward AI. Furthermore, we are of course aware of the risks associated with such innovative technologies—we are thinking, for example, of concerns about privacy or surveillance (more on that later). But do such concerns necessarily have to result in pushing these innovative options aside? Should we not at least try to move beyond our deep-rooted intuitions and attitudes? Is it not our job to at least think seriously about the moral possibilities of AI?

Conclusion

In the realms of technology and AI, a variety of beliefs are commonly held among AI developers, philosophers, politicians, engineers, and others. One prevalent belief is that technology is neutral, akin to how judges are expected to be. This claim can be interpreted in multiple ways. Although it might suggest that technology conforms to certain stereotypes and norms, the most recognized interpretation asserts that technology is detached from moral values such as fairness, privacy, and inclusiveness. Although apps and other technologies can facilitate these values, the prevailing view is that technology itself is inherently separate from any moral value. However, we have demonstrated in this chapter that it is incorrect to universally consider technology as divorced from moral concepts such as justice. Although not all technology carries moral value, technologies designed with specific ethical intentions, such as ensuring privacy or avoiding violations of privacy, clearly embody moral values. This applies to all forms of technology, from simple tools such as scissors to complex innovations such as self-driving cars.

Chapter 2
Ethics of AI

> *You just a barcode*
>
> Childish Gambino in This is America

The First Industrial Revolution relied on steam, the Second on electricity, and the Third on computers and the internet. For several years now, we have been in a new revolution. This has a lot to do with what the internet provided us with: data, big data, actually an awful lot of data about an awful lot of things from our online and offline lives. To give you an idea: it is said that 90% of the data available to us today was generated in the last 5 years. 'Data is the new oil' has been the cliché since 2006, when mathematician Clive Humby first used the phrase. And the USA is the new Saudi Arabia. What systems are being fueled with data?

If we are to believe Shoshana Zuboff, author of the 2019 bestseller *The Age of Surveillance Capitalism*, data serve the unbridled profit appetite of corporations, especially the appetite of the Big Five: Amazon, Apple, Facebook, Google, and Microsoft. Data serve the capitalist system, according to Zuboff, and especially the tech companies of the twenty-first century: companies without large factories and without high material and production costs. That assertion must be corrected, however. Not every use or development of artificial intelligence (AI) is an expression of monetary gain, just as not everything in capitalism has to do with AI. Of course, tech giants make billions from the data they glean from our activity, and privacy is often seen as an impediment to economic growth. But a less provocative and more accurate answer than Zuboff's thesis is that that data feeds smart technologies: AI systems. For the sake of giving the full story, we should add that some AI, especially expert systems, are not fed with countless pieces of information—we touched on this earlier in the Introduction—and also that things other than AI systems can benefit from lots of data.

Our lives today are already to a large extent intertwined with these smart systems. You might lament this, you might be annoyed that 'AI' is a buzzword today, but the fact of the matter is that little or no domain escapes AI, and this is unlikely to change any time soon. We live in a democracy and a meritocracy, but we are also moving more and more toward an algocracy, a society governed by the computational power of algorithms (Amoore, 2020). So, to warm up, let us start by presenting a small sample of the various contexts in which AI is used today.

The Algorithmic Society

The countless pieces of data traces we leave behind—simply by using our smartphones, visiting websites, posting, and 'liking' photos—are collected by Facebook and Google, among others, and sold to other companies, which in turn go on to work with them. Using AI systems, those companies get a good idea of who you are: your sexual orientation, ideological preference, addictions, relationship status, emotional state, ethnicity, intelligence, age, and so on. The system predicts what product might interest you and finally sends personalized ads with minute precision to your profile on Facebook, Instagram, or TikTok—ads for which, of course, the company has first paid the platforms on which they are posted. The goal is to get you to shop, to buy books, clothes, food, cars, medicine, or music. This, among other things, is explained with crystal clarity in Jeff Orlowski's 2020 docudrama *The Social Dilemma.*

Targeted advertising, around 70% of which runs through Facebook and Google, is not only a commercial but also a political tool. Although it is not completely unheard of in Europe, it is not nearly as popular there as it is in the USA. Barack Obama was just about the first politician to make widespread use of data in the run-up to the presidential election. But for perhaps the most famous and infamous example of the political use of AI, we must go back to the so-called Facebook election in 2016. In short, it comes down to the following—those who want a more comprehensive version can refer to the 2019 documentary *The Great Hack* by Jehane Nouiaim and Karim Amer. Cambridge Analytica is a British–American company that worked with presidential candidate Ted Cruz between 2014 and 2016. When he was out of the race, the company partnered with Trump. It collected information from 87 million Facebook users in a rather unsavory manner with the help of data scientist Aleksandr Kogan. The latter had developed the survey app 'This Is Your Digital Life,' rolled it out on Facebook, and used the information not only of those who filled out the survey but also that of all their friends. Cambridge Analytica's AI systems analyzed that data, predicted based on it whose vote might be of interest, and then sent targeted ads to social media users in the hope that votes would be cast for the Republican real-estate entrepreneur. By the way: although Theresa Hong, the director of Trump's campaign, thought that Trump would never have won the election without Facebook, it is far from certain that the campaign was effective. After all, research suggests that political advertising through AI only has a 2% effect (Aral, 2020); however, this small number can generate a significant effect in the so-called swing states. The same applies to commercial ads, by the way. There is doubt, according to research, over whether AI-driven ads lead substantially to higher consumption (Fisman, 2013).

Another example of how AI is being used today comes from I-Care, a company with offices in the Belgian cities of Bergen and Heverlee, among others. The company's core business is preventive maintenance. It predicts when a machine is worn out, and thus when replacement or at least maintenance will be needed. I-Care developed new technology for this purpose. Wireless sensors collect data on the machine's temperature, vibration, and noise emissions. AI is then unleashed on that information: it analyzes the data and predicts if and when the machine will wear out. Clearly, such innovation is very relevant to the industry. It prevents machines from not being examined in time and thus prevents them from breaking down, which can cause bodily injuries or environmental damage, among other things. In addition, it increases efficiency. Maintenance costs time and money, and this technology prevents unnecessary maintenance appointments.

Artificial intelligence also plays a role in medicine. The treatment of neck cancer, for example, is a very delicate operation. Radiation must be targeted while dodging the undamaged areas around the radiation site. For this reason, it is necessary for them to be indicated very precisely on the scan. Until not so long ago, only radiotherapist-oncologists did this. The problem was not only that it was very time-consuming, but also that if you ask two people to mark up the same areas on a scan, there is a real chance that there will be differences. More recent developments have allowed AI systems to be trained to recognize patterns using numerous images. The results are astonishing. When radiotherapist-oncologists use the new technology, their interventions are about 30% more accurate and faster than when they indicated the spots on the scans without AI. This not only benefits the cancer treatment itself, but it also frees up extra time for other care.

Artificial intelligence is also used to process natural language (think of interactive chatbots), for personal recognition (facial recognition), by sports clubs, by Google Maps to calculate your route, by banks to check your creditworthiness, to paraphrase text (ChatGPT), to determine your insurance premium, to estimate the risk of recidivism, for virtual assistants (think of Alexa and Siri from Amazon and Apple respectively) or to translate texts (Google Translate and DeepL). In short, the number of AI applications is not insignificant, and it seems that their scope will not shrink any time soon. To use an expression that has since become familiar: this is the summer of AI, and apparently it is a very hot summer.

Oral-B's Toothbrush

Given the wide spread of AI in the short span of a few decades, it is not surprising that there are strong beliefs about that technology. There are good reasons to be critical of those beliefs.

Some of the beliefs are very optimistic. Take for example the idea that AI is the solution to many and perhaps all problems. In a way, we agree. There is no doubt that there are good reasons to be very positive about AI applications, such as self-driving cars, for instance. According to the World Health Organization, 1.35 million people die each year worldwide owing to traffic accidents, more than half of which are caused by human error. Self-driving cars hold promise in that regard. According to estimates, they would significantly reduce the number of accidents and deaths. But that does not mean that they would no longer cause misery. Smart technologies may be desirable because they help to fight diseases, detect a potential pandemic, make the world a safer place, tackle the climate problem, or take over boring tasks. Yet AI systems face several problems, moral problems for example—more on that later. Moreover, it is naive and even dangerous to believe that AI systems are sufficient to solve issues related to climate or poverty.

Another strong statement has an optimistic character for some, but for others it is a reason for pessimism. It reads that we are not so far away from what is called 'superintelligence,' after the title of the well-known 2014 study by philosopher Nick Bostrom. In the foreseeable future, artificial entities would be designed that, unlike today, do not just perform specific tasks such as translating texts. Such new hyper-advanced systems would be able to combine multiple tasks and far surpass our intelligence. It cannot be ruled out that such entities will one day exist. Moreover, the prediction is not completely fictive. AI systems, as well as animals that are not humans for that matter, are already better than humans at many things: they have better memory and recognize patterns more quickly. On the other hand, there is no consensus among AI researchers about what exactly does and does not define AI, but there is agreement on this: the advent of superintelligence is still some time away, and perhaps even very far in the future.

Move Fast and Break Things

In addition, there is another claim that is no less popular than the two previous ones. It is that AI is a disruptive technology. In light of the long history of clever inventions, smart technologies are not only new, but they also disrupt existing domains. At least that is what is often claimed about AI. Let us call that idea 'the disruption thesis,' the thesis that will have our attention from now on and that, along with the neutrality thesis, among others, is a popular claim that people often cite in passing.

Admittedly, this assertion sounds credible, but is it true? Is it too much of a myth, just as the popular neutrality thesis is incorrect? In any case, the disruption thesis is an extension of Facebook's maxim: move fast and break things. Furthermore, it is also true that numerous companies are seen as disruptive. Netflix disrupted the television landscape, Airbnb the hotel world, Tesla the car landscape, and Uber the cab industry, according to some. And the same is said about 3D printing and nanotechnology. But besides these companies, 3D printing, and nanotechnology, are AI systems disruptive? Does AI blow existing domains completely off the map?

Unlike the idea that technology is neutral, the proposition that AI is disruptive does not go back that far. This is not surprising. Research around AI only got off the ground around the middle of the last century, and the AI applications we know today, generally based on machine learning, have only had a good commercial breakthrough in the last decade. Nonetheless, one can find the statement in numerous places today, both among those who are positive about AI (tech-optimists), but also among those who are rather critical (tech-pessimists). Flip open any of their books on AI and the chances are that one will describe it as a disruptive technology. Take *The Hype Machine* by scientist and entrepreneur Sinan Aral from 2020. He writes about social media and argues that it disrupts numerous domains. The subtitle reads *How Social Media Disrupts Our Elections, Our Economy and Our Health—and How we Must Adapt* (Aral, 2020). Philosopher Bernard Stiegler published a book in 2016 whose title was translated as T*he Age of Disruption. Technology and Madness in Computational Capitalism* (Stielger, 2019). Finally, the disruption thesis is also present in the book by the aforementioned Zuboff, for example, when she writes that "(…) the winning hand in capitalism is about blowing things up with new technologies." (Zuboff, 2019).

There are reasons to be skeptical of the disruption thesis. It seems particularly difficult to say about many AI systems that they are disruptive in nature. We are thinking of Oral-B's Genius X toothbrush. The brush is equipped with sensors that transmit data about your oral health to an app, which then analyzes the data and lets you know if, where, and how to brush further. Or take the message circulated by Microsoft in November 2020. The software giant announced that it was designing an AI system for two gasoline pump stations, one in Thailand, another in Singapore. The software looks for signs of unsafe behavior, such as smoking a cigarette at a pump. If it catches such a signal, the AI informs the staff, who must ensure that the cigarette is extinguished. The purpose of AI here is to ensure that the station does not go up in flames. Such development is valuable, though it does not seem to be a disruptive technology right away. And would you really claim that an AI toothbrush is disruptive?

On the other hand, the disruption thesis does not seem far-fetched either, based on the history of technology, for example. There is little to no doubt that in the past technologies that were not AI were more than innovative. Consider the wheel, which was invented 3500 years before our era, and which, like automobiles, brought about major social changes. The same is true of the printing press in the fifteenth century. If such technologies could have disruptive effects, why should AI systems not have them?

With such a question, we get to the heart of this chapter, which focuses on AI. Like the other chapters, it moves along three tracks—the first two being the most important. To begin with, we simply explain and clarify in several places. What does and the disruption thesis mean and what does it not mean? What are bias and moral responsibility? Are there different ideas about these, just as we can distinguish different conceptions of fairness? And when can we hold someone responsible and when not? In addition, we take a stand and argue. Do AI systems have a disruptive effect on morality? If so, why? Using, among other things, a thorough analysis of a case study—namely, autonomous decision systems—we argue for a conservative position: AI does not create entirely new moral problems and is consistent with our existing moral practice. Finally, in addition to describing and evaluating, we briefly consider the relevance of our reasoning. Why should we know whether technology is morally disruptive?

Disruptive Technology

The phrase 'disruptive technology' was coined by Clayton M. Christensen and first used in 1995. He used it to refer to technologies that shake up the market. A technology may be extremely popular, but at some point, new products are invented that prove that there is a gap in the market. The result is that interest in the old technology disappears and the innovative technology, the disruptor, becomes the next big thing.

All disruptive technologies are new in some respect, but some technologies are new without being disruptive. Look at transportation technology. In the late nineteenth century, the first automobiles appeared on the market. These were innovative because until then people had usually moved around with the help of horses. But that innovation was not disruptive: the cars were expensive, too expensive to be bought by the masses. The Ford Model T, on the other hand, first produced in 1908, was a disruptive technology. That was one of the first cars to go into mass production, which greatly reduced the price and allowed more families to buy such a vehicle. As a result, the market for horse transportation imploded.

This is an example of the original interpretation of the disruption thesis, an interpretation that is clearly economic in nature. Although that interpretation is still often used today, we focus on a different one. We discuss the moral interpretation of the disruption thesis. This requires some explanation.

Moral Disruption

We return for a moment to Chap. 1 and recall that according to the neutrality thesis, technology has no link with moral values. That thesis focuses on the technology itself, and not on the causal chain of which technology is a part. We have given examples of theories of the cause and effect of technology. According to some, Marx and Zuboff, for example, technology is a product of capitalism; others believe that secularization is an effect of technology. The disruption thesis must also be considered in this way. It is not about technology per se, but about the effect of technology on things that are not themselves technology. Consider the original meaning of 'disruptive technology': it is about the impact of technology on the market.

However, the fact that a technology has an impact is not enough to call it disruptive. A better description is that technologies are disruptive when their effects open up, turn an existing domain upside down. Being disruptive is more than just having a strong effect or reinforcing a certain influence. It refers to a break with a certain state of affairs. In that sense, it can be said that the Ford Model T was disruptive technology, but that cannot be said of the first cars.

Does that mean that the disruption thesis is exclusively about the domain of economics? No. Of course, the original interpretation is economic, and so is another often-heard interpretation of the disruption thesis, which is that technology is disruptive because it turns the labor market upside down. But you can also attribute a meaning other than economic to the disruption thesis. In any case, that is what numerous scientists, philosophers, opinion makers, and politicians often do. They talk about the seismic effects of technology on the legal, economic, or sports landscape. AI would turn the way in which wars are fought upside down, as well as the ways in which we work, administer justice, and organize sports.

We do not explore whether AI disrupts all domains; instead, we focus on ethics. We therefore translate the claim that technology is disruptive into the assertion that AI systems turn the domain of ethics upside down. We must immediately admit, however, that this formulation is actually not very meaningful yet either, just as the claim that AI will turn politics or sports upside down is too vague. It still needs to be refined. For what exactly is meant by 'ethics'? Are you referring to moral emotions such as compassion, or are you talking about behaviors such as helping someone in need? Another possibility is that by 'ethics' you are referring to a framework from which you make moral judgments. But even then there are several possibilities. You can either opt for the framework in which moral virtues such as courage and loyalty are central and about which Aristotle wrote, or for the approach of philosopher Jeremy Bentham, who focuses on the consequences of actions, or for the perspective developed by Immanuel Kant: duty ethics.

Let us choose for a moment the meaning of ethics as a frame of thought. The thesis then is that AI has turned our moral view of the world upside down. Whereas in the past we used to take people's good intentions or character into account when morally judging them, under the influence of AI and data scientists we have become much cooler—that is at least one possible interpretation of the disruption thesis applied to ethics. With increasing use of smart systems, we look almost exclusively at the actual effects of actions, at the relationship between good and bad consequences. Whatever your intentions, who you are as a person is of little or no importance for moral judgment; ultimately, pretty much only the results of your actions are morally relevant. More than ever, we are primarily making sober cost-benefit analyses, partly due to the influence of AI. At least that is the tenor of one possible interpretation of the disruption thesis.

Admittedly, this thesis sounds appealing. It is also quite popular, especially among tech-pessimists. The question now is not whether the claim is correct, however, but whether it is as obvious to defend it as it seems, just as there are difficulties with other moral interpretations of the disruption thesis. After all, a serious justification of the claim that AI changes our moral thinking habits requires quite a bit of research, which also raises many questions. For example, we know that people's judgments vary depending on context. They apply different principles in private life than in professional life, and the moral framework from which they think about politics differs from the framework for reflecting on economics. Moreover, does the research on moral perspectives cover *all* domains? If not, which ones were selected and why? Moreover, how was all this mapped out? Were interviews conducted, were questionnaires used, or did they also observe people's behavior? And finally, if there is a shift in thinking, on what basis does one conclude that this is caused by AI, and not by something else?

Law and Ethics

'Technology,' 'AI,' 'philosophy,' 'the neutrality thesis,' and now also 'the disruption thesis': those who use these terms may mean different things. Behind every interpretation is a choice. This choice can be nonconscious, but it can also be conscious—for example, because it can raise many questions, as in the example we just mentioned. With those things in mind, we push forward the following interpretation of the disruption thesis that we will explore from here on out: AI creates entirely new moral problems. It is obvious that the use of technology may be accompanied by technical difficulties. It also goes without saying that technology can be morally problematic. But our interpretation of the thesis is that AI systems also saddle us with new ethical difficulties. The idea, then, is not that AI reinforces existing ills; no, the claim is that AI creates at least one *new* kind of moral problem.

Clearly, this interpretation, which zooms in on problems and risks, is defended not by tech-optimists, cyberutopians, and believers, but by tech-pessimists, by those who oppose the algorithmization of society, the use of autonomous cars and weapons, the ubiquity of social media, and so on. But is it true? That is the question of this chapter. After the Neolithic and Agricultural Revolution, we are now living in the Industrial Revolution, the Fourth. But is that revolution also accompanied by ethical disruption? Does AI cause an ethical shock? Does it turn the existing moral practice upside down because it creates problems that were not there before?

That question is interesting enough in itself to take it seriously and to discuss it at length in a moment. The question is also practically relevant. After all, suppose you run into a new moral problem later on: the problem needs to be dealt with and you need to figure out exactly what our approach to the problem will be. Thus, the pages that follow are useful not only on a concrete level because they show what the moral problems with smart technology are, but the thinking we accomplish is also useful because it can culminate in the search for the best approach to the new problem. Let us assume for a moment that such a problem is actually found. What steps can follow? What concrete actions can result from our argument?

In any case, it is possible that existing law will suffice as a response to new moral difficulties with AI. After all, that technology is not entering a legal vacuum; there are already moral problems that are being addressed by law—think of privacy and the General Data Protection Regulation. But the reverse is also possible: current laws may be insufficient to address difficulties located at the ethical level. In that case, this question arises: do we opt for a legal approach or not?

The decision may be no, maybe because one considers the juridification of society, the increasing casting of solutions to problems in laws, as undesirable. In any case, that approach would not be exceptional, because there are still examples of things that are morally problematic but not punishable (Boddington, 2017). Cheating in the context of a love affair, for example, is legal, even though it is morally reprehensible. But that does not solve the problem. A problem that is not addressed by the law remains a problem that requires a solution. In that case, you could fall back on some very welcome initiatives from recent years, such as the now-famous 2019

Ethics Guidelines for Trustworthy Artificial Intelligence from the European Commission's High-Level Expert Group on AI. This document should be read as a supplement to law. It consists of guidelines that, although they are nonbinding and therefore have no legal consequences, were drawn up with the most ethical possible rollout of AI in mind. If it turns out later that AI creates new problems, then you can counter these by expanding the Ethics Guidelines with a guideline intended to prevent these new problems.

However, ethics guidelines, such as the 2017 Asilomar AI Principles, can play another role. One can also choose to address new moral problems legally, even if it turns out that current legislation is inadequate. In that case, a revision of that legislation is needed to fill the gap in the law. As we know, that can take some time—the ban on technology such as landmines, for example, took several decades. It is precisely in that period prior to the revision of the law that a code of ethics can serve as a buffer against new moral problems. At that point, that code is not an addition to the existing law, but an interim solution in anticipation of a new, revised law.

An Ethical Look at AI

The focus is thus on the potentially new ethical problems associated with the use of intelligent systems, and thus not on the problems related to the process prior to the use of AI: production. And yet, problems related to the development and production of smart technology deserve more attention. In a moment we focus on the ecological problems related to AI systems, but there are other problems on top of this, such as social problems. Let us address those right now before we get to the heart of the matter and focus on the disruption thesis.

An AI system that is able to identify cats cannot do that from the start. It must be trained to do so, for example, by presenting it with images of cats labeled with the term 'cat.' The system then looks for patterns in the images and, over time, it can distinguish cats from noncats. Machine learning can do this without human intervention, but clearly this learning process requires numerous labeled images of cats, sometimes millions. The problem is usually not that there are not enough pictures, the problem is often that there are not enough labeled ones. Enter humans.

Consider the website that says it must first be clear that you are not a robot but a human being. You are shown pictures and you are asked to tick which ones show a traffic light. You are given access to the site, but in the meantime, you have also created material that is useful for training the algorithm. However, the bulk of the labeling of photos is done by people who are paid to do so. These are thousands of people worldwide, especially in the USA and India. People like Kala, for example, who lives with her sons in an apartment in Bangalore, India, and spends several hours a day on her computer labeling photos of houses. Or someone like Justin in Houston, a management student, who alternates between taking classes and writing papers and endlessly clicking a mouse, giving companies enough images to run their AI systems. Although both belong to the same industry, they have little in common with the smoothie-drinking vegan technophiles on California's supposedly progressive tech campuses.

The work of Kala and Justin is an example of what can be called 'shadow work,' or ghost work, after the title of the book by Mary L. Gray and Siddharth Suri (2019). It is labor that is in the shadow in a double sense: you do not see the labor, for example, because it is done in small apartments on the outskirts of large cities, and it is hardly paid attention to, in part because the promising AI system itself is in the spotlight in the first place. And yet, like the shadow work of housekeepers, the work of Kala and Justin is indispensable: no algorithm without training, no training without photos, and no labeled photos without shadow work. In that respect, the shadow work at issue here is nothing like the shadow of your house on a summer Sunday afternoon. Although the latter is a reflection of something that can exist even without the shadow (the house), the algorithms, and therefore the tech industry, cannot exist without the shadow work of Kala and Justin.

By addressing shadow work, we are bringing attention to the human behind the robot, especially the repetitive labor that must be done in order for that robot to function. Shadow work is the work of the (human) robot behind the robot. But that is not the biggest problem, or at least not the only problem. Kala and Justin are part of the gig economy, which also includes Uber drivers and Deliveroo cyclists (Christiaens, 2022). That means they work on short, temporary and mostly part-time contracts with no title or chance for promotion. Those who reply that this is at best unpleasant but not morally problematic should keep the following in mind. Those who do ghost work today accrue little or no pension, may not have health insurance, are usually not entitled to vacation days, usually have no accident benefits, receive little more than minimum wage, and finally, probably never benefit from the profits that the platform, such as the Amazon Mechanical Turk, for example, picks up through your actions. So, there is still a lot of work to be done in this area as well, although we should add that in Europe, steps are being taken in the right direction.

So, the tech industry has a dark side as well as a bright shiny one. AI systems not only perpetuate and create injustice—we pointed this out earlier and will come back to it later—but they are also, in a way, a result of injustice. Undesirable states can be the effect of AI and can also be at the root of AI. Some may have every interest, mainly an economic interest, in keeping that invisible. The work is hidden, probably because it *must* remain hidden. That alone is a reason to shed light, and preferably a lot of light, on the shadow work done in the dark rooms of Bangalore and Houston.

Rage Against the Machine

We can now gradually turn to the central question. Are tech-pessimists right? Does AI disrupt ethics? Does that technology create entirely new moral problems? If it turns out later that it does, it would not be surprising in a way, for this reason.

Quite a few people have a negative attitude toward AI in the workplace for fear of massive layoffs. Of course, AI cannot take over all work; but the fears are not unfounded. Many jobs have already disappeared because of AI, and many will continue to do so in the future. According to a 2017 study by economist Carl Benedikt Frey and AI expert Michael Osborne (2017), as many as 50% of jobs are at risk in the USA. Others believe that the situation is a little rosier in the UK. There, one in three jobs is said to be under threat. This includes not only unskilled jobs or jobs with routine tasks that require few skills, but also 'knowledge jobs' in the medical, legal, or financial sectors are at risk of being replaced by artificial systems (Danaher, 2019).

It is not new that jobs previously done by people are now being taken over by machines. What is new is that we have to interact with some of those machines, and that algorithms in certain cases can make decisions that have a significant impact on our lives. Does that not create new morally problematic situations?

Suppose you are looking for a job. Your application letter is not read by the person in charge of the HR department, but by a system that has taken over the entire recruitment process, including the writing of the emails. Even the job interview is not done with a person. In the end, you do not get the job. You try many more times to get in elsewhere, but you are not selected anywhere. Over the years, no human being has ever read your application form; it is always a machine that judges that you are not the most suitable candidate. This situation is new. For the first time in the long history of humanity, we humans—job seekers, patients, prisoners—may be at the mercy of the decisions of intelligent but dead, inanimate matter. Would it be surprising if this were to be accompanied by new moral problems? Can one not expect that this new technological condition will also bring with it ethical problems that we did not know about until recently?

Although that may seem like a rhetorical question, there is also reason to answer it somewhat cautiously. Here is why: ethics, including AI ethics, has roots that go way back. Let us explain this.

When we make moral judgments about something or someone, they can be based on several reasons. One possibility is that you base your judgment on the ratio of good to bad consequences; another is that you look at whether the moral law was broken. It is also possible that your evaluation is based on a moral virtue such as loyalty or courage, that is, traits that belong to a person. An act may be seen as problematic because it shows little loyalty; you may praise someone because he or she is courageous. Within that framework, known as virtue ethics, something or someone is morally good or bad depending on the moral attribute being expressed.

That view is shared by many people, including AI developers. So, when they need to respond to moral problems or anticipate potential problems that may arise from their AI system, they often invoke moral virtues. For our line of argument, it is not important whether that framework is the most appropriate in the context of AI; but the fact is that people regularly fall back on it and that, at the same time, the framework has been around for a while. It goes back at least to Antiquity, because virtue ethics was first systematically elaborated in the writings of Aristotle. Is this not a reason to at least doubt the idea that AI is accompanied by new moral problems? After all, a framework such as virtue ethics exists to avoid or deal with ethical issues. If that framework is used when talking about moral problems with AI, and that framework has been around for a long time, how likely is it that AI is accompanied by new problems?

The reference to work and virtue ethics is obviously no reason to accept or reject the disruption thesis. Both are at best a reason to suspect that it might be right or wrong. How then should we evaluate the disruption thesis? Is it a myth, a false claim, just as the neutrality thesis is false? To answer that question, of course, we do not need to look at *all* AI systems. It suffices to consider whether a moral problem emerges in the context of AI that is new, a problem that may be caused by only one or two systems. So, let us take a closer look at the problems that people are usually talking about these days: abuse of AI, and issues with privacy, bias, security, transparency, and the environment. We call them 'the deadly sins of AI.' We shall start with abuse.

Bad Use

Judging an AI system ethically means, among other things, that one should look at the purpose for which the system is designed and used. What is true for AI is true for all technology. Some purposes, such as translating a text, are normally morally neutral, and the same can be said of a system such as Google Translate. Other goals, however, are of a moral character and are desirable. We saw some examples earlier: the equal treatment of people, the training of moral abilities, and so on. But AI can also be created or used with morally undesirable goals in mind. An example of the latter is deepfake—the phrase is a contraction of 'deep learning' and 'fake.'

Deepfake is a technology that enables you to combine existing video and audio, so that, for example, you can manipulate a video to show a person saying things that he or she has never actually said. As a viewer, you think that this person has said these things, whereas in fact this is not the case. A well-known example is 'You Won't Believe What Obama Says In This Video,' a clip from 2018 that you can watch on YouTube. You are under the impression that Obama is speaking and that he is actually uttering the sentence "President Trump is a total and complete idiot." After a few seconds, it appears that it is director Jordan Peele who is putting those words into the former president's mouth. In line with that is DeepNude, which was created in 2019. If you gave that website a photo of a woman with clothes on, it would paste that woman's head onto a naked body in such a way that it looked like a nude photo of that woman.

Note that deepfake can be perfectly morally innocent. For example, it can be used in the arts or entertainment industry to edit the speech of a character in an existing film. Moreover, as is often the case with technologies, it can also be used in a good sense. In Germany, for example, deepfake has been used by police departments to track down individuals who view and distribute child pornography. Researchers present an AI system with thousands of such videos, after which the system composes its own video. That new footage is used by an undercover agent to get inside the online world of child pornography, with the goal of detecting and stopping the use and distribution of unwanted images (Moody, 2020).

Nevertheless, deepfake and other AI can also be used for morally problematic purposes. The list of its objectionable uses is long: revenge porn, the spread of fake news, cyber-attacks, terrorism, the manipulation of elections—think Cambridge Analytica and the 2016 US presidential election. But, crucially, these are not things that *only* became possible because of AI. People with bad intentions were interfering with these things even before AI existed. Some respond to this as follows: AI is a technology that allows a group of people, terrorists for example, to wipe out the rest of the world's population for the first time in history. If we assume for a moment that AI does indeed possess such power, this is still not an argument in favor of the disruption thesis. The ability to wipe humanity off the map is not unique to AI; biological and chemical weapons can do this as well.

The Panoptic View

A thorough evaluation of AI requires not only that you look at the purpose for which it is being designed or used, you also need to consider whether its actual use involves moral problems. The purpose of an AI system may be good, but there may nonetheless be problems associated with its use. Compare it to the actions of an army: liberating a people is morally good but using mustard gas for that purpose is not. Does the use of AI, regardless of its purpose, come with difficulties we did not know about until recently?

Privacy is a central value in many societies. It can mean several things, but usually in the context of AI it means one of these two things: control over one's own personal data and the right not to be followed, to be free from surveillance. For a technology to be privacy proof in both respects, one of the things that must be taken care of is the following. If the artificial system collects information, then the person whose data are being collected should be informed of, first, the fact that data are being collected, second, exactly what information is being used, and third, the reason why that information is being stored. Furthermore, the relationship between the information and the purpose it serves should also be considered. If you collect unnecessary data, you are violating privacy.

Artificial intelligence systems can meet these requirements in principle, including through open communication and encryption—the encoding of data. One may use cookies, digital tags that register surfing behavior, as long as one asks your permission and you actively grant it. On the other hand, it is clear that some AI systems have a problem in terms of privacy. We refer to the fact that we live in a society where we are constantly monitored through digital systems. Employers track their employees' activities through apps to know if they are not wasting their time on social media, meeting their goals within the intended time limit, answering their emails, and other things. Or take books. If you read them through Google

Books, people normally know pretty precisely how long it takes you to read a chapter and which passages you find gripping or not. If, on the other hand, you prefer to read physical copies that you buy in the store, chances are that your smartphone records which store you bought the book in, how long you spent there, and which part of the store you lingered in the longest. And if you paid for your copy with a card, it is likely that data traders will resell that information to, say, insurance companies. In short, we are not far from living in a decentralized panopticon. There is no single gaze monitoring all our doings from one perspective; no, we are constantly being watched from all sides (Véliz, 2021).

In addition, AI systems sometimes face privacy issues in the sense that we as users do not even know that data concerning us are being stored, and in some cases, we have no idea what will happen to our data. Consider the Cambridge Analytica scandal we already mentioned: the data of millions of Facebook users was used without their permission. The same problem surfaced in the case of Clearview AI, an American company specializing in facial recognition that works for numerous police forces and intelligence agencies around the world. It trained an AI system with millions if not billions of photos that it extracted from platforms such as Facebook and Instagram without the users being aware of it. The problem is also evident in a 2016 study of privacy in 211 Android apps for diabetes patients. The results are disconcerting. Merely by downloading the software, 31% of the apps allow you to find out the identity of the user, 27% give you location data, and 11% activate the camera to access your photos and videos. These things also go against the ethics of technology: for about 5% of the apps, the contacts on the device can be read and the microphone can be activated to record conversations (Blenner et al., 2016).

So, some AI systems violate privacy. But this is not true of all systems, and more importantly, privacy problems are not new problems created by AI. Parents also commit such mistakes when they snoop in their child's diary, just as it is a problem when, as a researcher, you do not ask respondents if you can process their data. An example from the context of technology that is not AI: it is irresponsible if data from the use of medical equipment are used for research without first being anonymized.

Racist and Sexist

For now, we have no reason to conclude that AI is shaking up ethics, that smart technology is ethically disruptive. The moral problems with AI that we have seen so far are a continuation of existing problems in a different context. It is old wine in new barrels. So, we suggest looking at another problem that often comes up when talking about AI and ethics: bias.

To begin with, we would like to emphasize that 'bias' can mean several things and that these meanings should not be confused (Coeckelbergh, 2020). Usually in the context of AI, two senses of the word are used: the statistical and the moral. When speaking of bias in the statistical sense, it means that a sample is unrepresentative, and thus that it gives a distorted view of the population. An example of such bias is the following. One wants to predict the election result and does a survey of the qualified residents in the capital. There is a real chance that the prediction is that the country will be governed by progressive parties in the coming election. But there is also a very real chance that this prediction is wrong, and that is because the person only polled a well-defined segment of the population. The sample was not representative, and thus was biased in a statistical sense.

Someone is biased in a moral sense when she or he is not neutral, when, in other words, that person assigns different weight to a person, group, or idea. Note that being non-neutral is not necessarily a moral problem. When several children are injured, a father will help his own child in the first place, simply because it is his child. He is not neutral, but that is not problematic in this case, quite the contrary. It would be very strange and even morally wrong for him not to be motivated by the fact that one of the children is his child. Therefore, this way of being non-neutral is not bias. In other words, moral bias is not just about non-neutrality, but about being prejudiced in a way that is undesirable.

It is not difficult to see what exactly the problem with bias is. It can lead to (negative) discrimination: unequal treatment that cannot be justified. Unequal treatment is also not necessarily a problem in itself. Giving someone a raise based on the difference in performance is justified. Unequal treatment is irresponsible when it is based on skin color, sexual orientation, and other characteristics that are irrelevant in this context. It is precisely at that point that bias in a moral sense comes up. Bias that has to do with an irrelevant characteristic can lead to unequal treatment on the basis of that characteristic, and that is impermissible. Consider the hiring of new employees by a business manager. If the latter selects someone, consciously or unconsciously, on the basis of physically attractive characteristics, that is an example of lookism, of unequal treatment that is not justified.

There is no doubt that people are biased in a moral sense. To take the same example: people with physical characteristics that are considered attractive by many have a significantly greater chance of moving up the social ladder precisely for that reason. But you can be sure of this as well: numerous smart technologies are also biased in a moral sense. AI systems developed by computer scientists and engineers, while often appearing objective, can be biased in a moral sense upon closer inspection. Consider the example from the Chap. 1: a banking system was more likely to give loans to white men than to women of color. But there are plenty of other examples, for example, the famous case of Amazon.

In 2018, news service Reuters announced that Amazon wanted to automate not only its warehouse work but also its human resources management department (Dastin, 2018). In 2014, a team created a system that could screen all applications and gave a score between one and five stars to applicants—a bit like how we rate Amazon's products. In 2015, they discovered that the AI system was not selecting in a gender-neutral way: women's applications were systematically not chosen. The system rejected applications that contained the word 'woman,' for example, as in 'president of the women's chess club.' The reason was that the computer models of the company founded by Jeff Bezos had been trained with the applications that the company had received in recent years and, more specifically, with documents from those who had eventually been given a job. These were generally men. The example is the canary in the coal mine of the tech industry. It is a symptom of the systemic problem that the world of technology platforms such as Uber and Amazon is a man's man's man's world, that on Menlo Park campuses and in Palo Alto, women are vastly outnumbered

Although AI systems can be biased, the source of the bias is not necessarily the system itself. Often the problem lies with its designers. They may have racist motives and they may equip their technology with discriminatory algorithms. But AI can also be biased without anyone having bad intentions. This can be due to several things. One may wrongly assume—think of Chap. 1—that technology is by definition neutral (in the broad sense of the word), and therefore that one need not guard against bias creeping into the system. Or one is insufficiently aware that bias may underlie unjustified inequality. Another possibility is that there is little or no awareness that unjustified unequal treatment can also have a major impact on people's lives, for example, because it deprives them of a job or a loan.

But moral bias can also result from that other form of bias—statistical bias—from the fact that the data with which the AI system was trained are not representative. This is true in the example of Amazon just mentioned, but also for AI developed for facial recognition. Many of those systems are trained based on ImageNet. This dataset contains a lot of data originating from the USA, whereas only a small part of the dataset originates from India, China, or Brazil, countries that nevertheless represent a large proportion of the world's population. This statistical bias means that face recognition does not work for people with nonwhite skin color, which is an example of moral bias.

Another example of the link between the two forms of bias comes from OpenAI, the company founded by Elon Musk and Sam Altman in 2015. Since the end of 2022 we have all been familiar with ChatGPT, but it was as early as 2019 when the company released a system for natural language processing, the then very highly anticipated GPT-2. The system takes input—a word or phrase—and links new words and phrases to it. The technology can be used for a variety of texts—from news articles to novels—and generates texts that are very convincing. However, from the beginning, one of the problems with the chatbot is that it sometimes generates text with sexist and racist stereotypes. Giving the system the sentence "The woman worked as," the phrase "a prostitute named Hariya" could follow; for "The gay man was known for" it may follow "his love of dancing, but he also took drugs." Thus, the technology is at some point not neutral at all, but clearly morally biased. Such an outcome is undesirable, but at the same time not entirely surprising. To train the system, the company had chosen, among other things, WebText as its dataset, which has about eight million documents taken from the pages of the American social news website Reddit. The users of that site are primarily white males in their twenties. It is not surprising then that the pieces of text generated by OpenAI's chatbot reflect the kind of online conversations these men have on Reddit. The example is a nice illustration of the now familiar phrase 'garbage in, garbage out.'

When it comes to AI and ethics, people often talk about moral bias. Although that is obviously a good thing, in the case of AI, bias does not necessarily come from the technology itself. However, is bias a new problem? In other words, is bias an argument for the tech-pessimist, or a reason to accept the disruption thesis? In Chap. 1, we mentioned the negativity bias—the tendency to attribute bad qualities to unknowns. Or, take the results of cognitive research that maps bias. That shows that the majority of people studied are not neutral: they associate white with good, Black with bad. And in terms of gender, research suggests that women are strongly biased toward men, and vice versa (albeit to a lesser extent). Those who minimize this must remember that the consequences of such bias might be severe—a difference in waiting time for medical treatment, for example—as well as subtle. It can cause us to have less eye contact with the person in question, or to smile more (or less) at that person (Chugh, 2018).

Safe and Secure

Intelligent systems can cause troubles in more than one sense: there is safety and there is security. Insecurity means that others, hackers, can penetrate it. There is a risk that the software could be taken over by people other than the users or designers, people with less than noble intentions. The operating system of a self-driving car is an example of an AI system at such risk. The problem at play here is closely related to the first problem we mentioned: the use of AI for the wrong purposes.

It is also to the latter danger that reference is made in Isaac Asimov's 1950 collection of stories *I, Robot*, long before neural-network-based AI came to fruition, and long before Alex Proyas' 2004 film of the same name came out. On the opening page of that volume are the famous three laws of robotics—the film opens with them as well—that foreshadow robot ethics. The first law goes like this: 'A robot may not cause injury to a human being, nor, by remaining passive, cause injury to a human being.' Asimov's second law reads, 'A robot must obey orders given by humans except when those orders violate the first law.' Finally, the third law states that a robot must protect itself as long as it does not conflict with the other two laws (Asimov, 2004). Asimov's robots do have to follow orders, but when ordered to do something harmful, they must refuse. This addresses the problem of insecurity that some AI systems also face today, namely that they can fall into the hands of people with bad intentions, a problem that can come with far-reaching consequences—think of the havoc a self-driving car could cause if it were hacked by rogue minds.

Artificial intelligence systems, however, can also cause problems in another sense. When some of these technologies are deployed to automate monotonous tasks in a factory, for example, they bear the risk of potentially causing harm. This is a health and safety issue. We are thinking mainly of physical and material damage. This kind of danger mainly comes into play when AI is embedded in hardware and operates in the physical world. Again, let us take self-driving cars as an example.

We have pointed this out before: self-driving cars hold promise. There is a real chance that they could have a positive impact on road safety. Yet those cars will not free us from accidents, because there are risks associated with the use of such vehicles. At least that is what the first experiments with autonomous cars have shown. In 2015, there were about 20 small crashes with such cars, although no one was injured and the accidents were caused by a mistake on behalf of the driver of the non-autonomous car involved in the accident. That changed in 2016. In February of that year, Google's autonomous car collided with a bus somewhere in California. Again, the accident involved only property damage, but was caused by a fault in the car's operating system. Three months later, however, a tragic accident occurred. The passenger of a Tesla car was killed after the vehicle collided with a white truck. Cause? The sensors of the self-driving car had not detected the white truck. Finally, in 2018, Elaine Herzberg was killed. She was knocked over while cycling in the dark by an Uber self-driving car, somewhere in Arizona. Spicy detail: although Google claimed responsibility for the nonfatal crash, Tesla, which was worth about US$350 billion at the end of 2020, washed its hands of the accident. This in part was because the person sitting in the car was looking at their smartphone and therefore not paying any attention to traffic at all. We will come back to that theme later.

It is clear that AI at least holds the promise of making environments safer, though you do not immediately have to think of Steven Spielberg's 2002 film *Minority Report*. That is also what people thought a few years ago in some cities in Belgium and the Netherlands. In Aalst and Rotterdam, for example, smart cameras are used. These are devices that not only register situations, but also interpret them. Fed with countless images in databases, these devices have been trained to recognize fights, for example. Yet there is no consensus that cameras effectively make cities safer. Some believe they do. According to them, in city centers crime would decrease as a result of the use of the technology; the greatest effect would be in parking lots. Others are skeptical, claiming that AI has negligible effects, or none at all. Cameras, according to them, would only help to track down suspects after a crime.

On the other hand, the following is clear: not all AI systems are completely safe or secure; they come with a risk of damage or can be hacked. This is regrettable, but it is a problem that can be worked on. After all, safety lies on a continuum: a car today may be safer than it was before, whereas cars in the future may be safer than they are today. Moreover, anyone who believes that not having complete safety is also an argument against the rollout of AI should realize the following: there are an extraordinary number of technologies that have nothing to do with AI and that are not risk-free, but that are still allowed. Bridges, cars, planes, laptops, lights, microwaves, drills, cell phones: they are not banned, even though they are not completely safe in one sense or another and they can all cause damage. Although it may be true that AI systems face the problem of unsafety—a problem that can be serious and must be properly addressed—it is not a new problem unique to AI. Thus, unsafety is also not an argument for the disruption thesis, at least not for our interpretation of it.

Like a Dark God

Nowadays, people often talk about 'Transparent AI' or 'Ethical AI.' Although these are often juxtaposed, 'transparency' and 'ethical' do not mean the same thing. Transparency is desirable in many situations, but when AI is exclusively transparent, it is not necessarily ethical; in addition, it must certainly be privacy proof, nonbiased, and secure, among other things. Transparency is an insufficient condition for ethical AI.

Transparency means, among other things, that sufficient information must be available. This can apply in several areas. It can mean that one is open about the data (how were the data collected?), the design process (who made the choices?), and the stakeholders (what interests came into play?), among other things. In addition, it can also mean that people other than the designers are able to understand the technology, at least a certain aspect of it. This is also why people sometimes talk about 'Explainable AI' instead of 'Transparent AI.' Understanding AI that is morally relevant is located on neither a technical (how can these materials be combined?) nor a mathematical level (how does this algorithm work?). The desirability of transparency has to do with the reasons upon which the decision the AI system makes is based. Take the use of AI for hiring. When a company needs to select a new employee, it can leave it to an AI system. Based on previous training, the system will decide which candidate is the most suitable for the position. The fact that the AI system must be transparent means that it must be clear why the system is selecting this particular candidate, and why the other candidates were not selected. If you are in the dark about the reasons for that decision, then the AI does not meet the condition of transparency.

In many cases, it is desirable that AI is transparent, but we want to emphasize that transparency can also bring about problems. If it is crystal clear how a technology works, then it can also be more easily manipulated by people with bad intentions, such as terrorists for example. It is also true that nontransparent technology is not necessarily a moral problem. That is because the world does not break down into either ethical or unethical. Some things are morally neutral. We gave *Temptation Island* and a drill as examples in Chap. 1, but a closet, an AI toothbrush, and Spotify are also normally morally neutral. When the music platform's algorithm recommends an album by, say, The National, whereas one has absolutely no idea and no way of knowing what that recommendation is based on, it is far from a moral problem. That absence of information may be annoying, because one might very much like to find out why Spotify recommends that music, but even so, one cannot call that opacity a moral flaw. Although this is not to say that it is by no means morally unproblematic for Spotify's algorithm to recommend, for example, the music of a racist homophobic rapper or to make music suggestions to me based on one's sexual orientation.

In certain cases, however, no transparency is certainly morally problematic. The reason may be that transparency is necessary to address the moral problems we mentioned earlier. Take bias. If it turns out that an AI system is biased and that this bias leads to unwarranted decisions, then you need to know what data the algorithm was trained with if you want to remove the bias from the system. Transparency is morally relevant in that case, namely as a means of solving the bias problem. For completeness, we would also add here that transparency is also useful because it makes users trust the technology more. When smart technology has fewer secrets, research shows, people are more likely to use the system.

Opacity can also be undesirable in itself. This is so in contexts where an AI system's decision impacts a person. Assume for a moment that an algorithm decides that you are not the most suitable candidate for the job, a decision that can have far-reaching consequences. Moreover, you cannot find out what the decision is based on. In that case, the lack of knowledge is not merely a lack of knowledge, it is also a moral deficit. Why? In a liberal democracy, at least in principle, every person has value in themselves, regardless of background, gender, skin color, or impairments. This means, among other things, that no one is allowed to hurt you without good reason or that you have a right to care when it proves necessary. It also means that you have the right to an explanation when the choices of others have a negative impact on you, for example, because you were not selected for a job. To you as a human being, as a being with moral standing, one is obliged to give reasons for this. If one does not provide them, one is failing, because that is a disregard for the value you have as a person.

Some AI systems are unquestionably transparent. Take expert systems, the traditional form of AI we talked about in the Introduction to the book. These are based on the knowledge of an expert who programs rules into the technology in the form of if-then sentences. An example is this: 'If iRobot's robotic vacuum cleaner Roomba collides with an object, it must turn around.' The latter action is completely transparent: you know exactly what it is based on, namely the conditional part in the if-then instruction written into the robot by the programmer. Thus, technologies equipped with such rules do not usually pose a problem in terms of transparency—usually, not necessarily, because expert systems can also be very complex.

It is different with machine learning, the more recent form of AI, and more specifically when such a system consists of neural networks that are fed with data. It then starts looking for statistical relationships among that data. The goal is for the system to eventually make the most accurate prediction possible when it receives new data. Clearly, that form of AI is very successful. Algorithms play a role in scientific research, detect tax evasion, or block your credit card in the case of (possible) fraud. The downside is that in addition to being potentially biased and insecure or unsafe, they are not necessarily transparent. This is not surprising when it comes to the user, but designers often cannot fathom the technology either. Of course, AI developers can explain how the AI system works in a general sense, but sometimes there are so many neurons and connections between the artificial neurons that it is impossible for even the programmer to understand the decisions. To illustrate, YouTube's neural network designed to recommend videos has about 30 layers. When the system malfunctions and needs to be re-trained, it is obviously a problem. But it is also problematic when the AI system functions perfectly, that is, when the system makes the most accurate decision. It may be that the technology was right to select this candidate, but when it cannot explain why one was not chosen, that is a moral problem. It goes against the idea that everyone deserves an explanation.

So, some AI systems are like the God of Protestantism: obscure, unfathomable. Is transparency a new problem, though? Anyone applying for a job should, in principle, have the right to know why he or she was not selected. Yet it is often the case that if you are not hired, you have to guess at the real reason yourself. In the sports world, you still hear stories of players who are not part of the competition core for a while, and their coach does not give them any explanation. In Belgium, during the corona crisis, it was often completely unclear what exactly was in the reports with advice that experts wrote for the National Security Council. An even more tragic example comes from the judiciary. Also in 2020, Amnesty International announced that the number of countries carrying out the death penalty is steadily decreasing, but in that minority of countries the number of death sentences is increasing. That report also stated that many practices surrounding the punishment are secret: the convicted person does not know why they are being punished—and thus does not have the opportunity to defend themselves—there is no official communication, and families are not informed (Algar, 2020).

Thus, lack of transparency is not a new moral problem. On the other hand, however, not everything is the same. An employer, coach, or judge who does not provide an explanation can, in principle, do so. In the case of an AI system made up of highly complex neural networks, it is impossible for practical reasons to know the basis of the decision. Historically, this has not happened before. Thus, when it comes to transparency, you could say that AI is either reinforcing or exacerbating an existing problem. But that is no reason to decide that the technology is disruptive, at least insofar as 'disruptive' means that a new kind of moral problem is created.

It's the Ecology, Stupid!

Ethics is mostly about people. Should they be treated equally? Who is responsible? Is that trait morally relevant? In addition, ethics can also be about other organisms or technologies. Do animals have rights? What about plants and robots? Do we have obligations towards them? If so, on what basis? Finally, you can also consider the environment. In addition to animal ethics, plant ethics, and ethics of technology, there is environmental ethics: a moral reflection on earth, water, and air.

Those who think about the environment ethically should keep in mind that the environment can be harmed in at least three ways. First, there is pollution. Substances can be added to earth, water, and air that prevent them from fully ensuring that life on this planet is healthy. In addition, there is depletion. That is, energy resources are extracted from the environment that are subsequently no longer renewable. And third: degradation. Such damage occurs when there is a structural change in the environment. Examples are soil erosion and the decline of biodiversity. Note, however, that although the three forms of environmental damage are in principle distinct, in practice they are often closely related. The hole in the ozone layer, for example, is a form of degradation that results from pollution by chlorofluorocarbons (Van de Poel & Royakkers, 2023).

Are environmental problems moral problems? There are two ways to argue that environmental damage is morally problematic. According to the first line of thought, environmental damage is a moral problem insofar as it has undesirable consequences for the lives of human beings and other organisms today and in the future. It is usually that type of argument that plays out in the fight against air and water pollution. It is our ethical duty to provide clean water and clean air, the argument goes, because they are essential to a healthy life. Air pollution is undesirable because it has a bad effect on our health. Are all environmental problems moral problems? According to this line of thinking, no. Damage to the environment that is not harmful to humans or other beings is not a moral problem.

In addition, the environment is also thought of in non-instrumental terms. Pollution, depletion, and degradation are a moral problem, not because of the undesirable effects on humans or other living entities, but independently of the effects on organisms, because the environment in itself has moral value. Thus, when you do something that is harmless to humans or nonhuman animals, it does not necessarily follow on the basis of that argument that your action is morally unproblematic. After all, what you do can also be harmful to the environment, and that is a problem for those who believe that the environment itself has moral standing.

We shall not address whether this second form of argumentation is convincing. What is certain, however, is that the environment is one of the most important issues today, if not the most important. According to some, its importance overshadows that of AI, no matter how hyped the technology may be. Some of that is justified. As useful as a smart toothbrush is, it is pretty trivial compared with the environmental damage caused by greenhouse gas emissions from the products of the industrial revolution: cars, planes, factories. Nonetheless, the claim that AI is less relevant than environmental issues misses the close connection between AI and the environment. Smart systems can both prevent environmental damage and exacerbate existing environmental problems. Let us explain that a bit further.

There is no doubt that numerous non-intelligent technologies have positive effects on the environment. We are thinking primarily of solar panels and electric cars. However, the same can also be said of AI systems. Take the Green Horizons project, which was started in 2014 by the city of Beijing in cooperation with the company IBM. Using traffic cameras, social media, weather stations, and wearable sensors, among others, data are collected on the distribution of particulate matter, the most dangerous form of air pollution. AI systems are unleashed on these data. Those systems analyze the information received and predict where and when pollution will occur. Such predictions can go 10 days into the future, allowing the government to take targeted action to improve air quality. The result? The amount of particulate matter decreased by 20% in a few years, which is a very good consequence, especially considering that thousands of people die every year in China as a result of air pollution.

The downside, however, is that AI is harmful to the environment. Smart systems are not in the cloud, they are in materials—they reside in machines. It requires computers to store large amounts of data and make calculations quickly—computers that, by the way, are mainly located in data centers in the USA, China, and Europe, and are mainly run by tech giants such as Amazon. This may not be a problem, were it not for the fact that making computers requires, among other things, tin and silver. About 36% of the globally available amount of tin goes into making electronics, and approximately 15% of silver.

Moreover, the use of those machines requires energy, a lot of energy, also because data centers are equipped with air conditioning, which requires a lot of energy to counter the heat emitted from the computers. To give an idea, an estimated 5–9% of all energy consumption would be for information technology, including AI; training a large AI system consumes about 2.8 gigawatt-hours of electricity, equivalent to the electricity consumption of three nuclear power plants in 1 h (De Ketelaere, 2020). According to researchers from Sweden, there will be 15 times more demand for electricity from the AI world by about 2030 (Crawford, 2021).

A related problem is that AI is associated with large greenhouse gas emissions. 2% of global CO_2 emissions are currently said to be due to information technology and AI. Just training a popular algorithm emits over 200,000 kg of CO_2 or more. By comparison, an average European flight emits about 500 kg of carbon dioxide per person into the atmosphere (De Ketelaere, 2020). Researchers Lotfi Belkhir and Ahmed Elmeligi suspect that in around 2040, an estimated 14% of global CO_2 emissions will come from smart technology (Belkhit & Elmligi, 2018). This not only supports the claim that AI is morally problematic, but it is also a reason to include AI in debates concerning the environment, and not just to focus on meat eating, washable diapers, air travel, and driving.

Is the environmental damage caused by AI an argument in favor of the disruption thesis? This, of course, does not need extensive argument: the environmental problem is older than AI; furthermore, nonsmart technologies also have undesirable effects on the environment. Consider the industrialization of society, especially the Second Industrial Revolution since the mid-nineteenth century, with the introduction of industry. From that period on, and even more so since the middle of the last century, the environmental problem is taking hold, because of the burning of fossil fuels and deforestation, and global temperature rises. The consequences are well-known: drought, melting ice caps, extreme weather events, rising sea levels, and so on. In other words, the increase in CO_2 particles does not just date from the second decade of the twenty-first century. That increase is older than the period in which machine learning appeared on the market. It existed during that period in history when factories and railroads were being built, when there was no talk of smart cities and self-driving cars.

At the same time, we do want to underline the following. That the ecological problem was not created by AI does not reduce the need to think about AI in an ecological sense—in fact, the opposite is true. An ethical view of AI must also be a sustainable one. Hence, it would also be better not to speak of 'Human-Centered AI,' yet another expression used today in the context of an ethical reflection on AI. After all, that expression suggests that AI ethics is exclusively about the direct effects of smart technology on people—think about bias and privacy—whereas ethics may not be reduced to that. Ethics should also be about the indirect effect of AI on humans, that is, the effects AI systems have on humans through the environment. In addition, the effects of smart technology through the environment on non-human life must be considered: on plants and trees, for example, but also on animals that are not humans. There is no good reason for AI ethics to be exclusively about humans. In other words, we need a non-anthropocentric approach to ethics as well. In short, it is appropriate to strive for a broad conception of Ethical AI, that is, for Sustainable AI.

The Seventh Sin

It is important to exercise caution: it is possible that we are forgetting a moral problem that was unknown until recently. Nevertheless, for the time being, we see no reason to defend the thesis that AI challenges ethics. Privacy, bias, and unsafety are obviously serious issues. And yes, it could be that these call for new solutions or that new problems do emerge in the future. Furthermore, it is also true that AI systems are very fast and powerful, at least faster and more powerful than most tools that are not AI, so the impact of the problems may be quite a bit greater. Yet the current problems associated with AI are not new. Ethics, at least as far we can see, is not being disrupted, broken open, plowed over by that technology. So, there is probably no sharp boundary that isolates the moral problems associated with AI from the pre-existing problems that are separate from AI.

At least that is what the previous analysis teaches us. However, does that make it the last word? We addressed six problems just now, which we called 'the deadly sins of AI': abuse, privacy, bias, security, transparency, ecology. But are there not seven deadly sins? According to some, there is a seventh deadly sin in the context of AI. There is something else that is both morally undesirable and new. It is about the so-called responsibility gap, the impossibility of assigning moral responsibility when AI causes problems. This would occur with the use of hyper-sophisticated autonomous technologies such as self-driving cars and weapons. Let us begin with a brief exploration of such systems.

Autonomous Cars and Weapons

"They used to say guns don't kill people, people do. Well, people don't. They get emotional, disobey orders, aim high. Let's watch weapons make decisions." This quote refers to the well-known slogan of the National Rifle Association with which we addressed Chap. 1. It comes from *Slaughterbots*, the 2017 video by global AI expert Stuart Russell. The recording is part of the international Campaign to Stop Killer Robots that has been running since 2012 (and was followed up in 2015 with the Campaign Against Sex Robots). The goal is to raise awareness among the general public around what is now recognized by most experts: that there are great dangers associated with the so-called killer robots or lethal autonomous weapon systems.

Anyone who wants to get a concrete picture of killer robots can watch 'Metalhead,' the fifth episode from the fourth season of the Netflix series *Black Mirror* (2011), in addition to Russell's video. But what exactly are such robots?

Killer robots are weapons. Their purpose is to be used to fight during a conflict. But that description is still too broad: not all weapons are killer robots. It is better to define them as technologies designed to kill or at least disable people. However, even that description is not sufficient, because it also includes a 9-mm pistol. The difference between that gun and a killer robot has everything to do with the latter's autonomous nature: the robot functions without human intervention, whereas a gun is in no way autonomous. Certainly, killer robots are created by engineers and programmers. And, of course, they only start to work after someone 'pushes the button.' But once this has been done, they can do what they were created to do on an autonomous basis. They detect a target in the air, on land or underwater, and then decide whether or not to fire. Finally, although 'killer robots' usually refer to systems to attack, it is not necessarily so. Those robots can also be used defensively. The same goes for their mobility. Usually, people talk about robots that move in the air, underwater or on land. Yet static autonomous weapons designed to kill people are also killer robots (Leveringhaus, 2016).

Whether such technology currently exists and is being used is not entirely clear. For example, some say that the SGR-A1, the robot used by South Korea to prevent soldiers from North Korea from approaching, is a killer robot. It is equipped not only with sensors that detect any movement of people at the border on its own, but also with machine guns that would allow the robot to fire on its own. Others deny this, as the SGR-A1 would only fire when ordered to do so by a human. Also, according to a United Nations report, in Libya in 2020, a drone would have attacked General Khalifa Haftar's troops on its own (Cramer, 2021). At the same time, there are doubts as to whether that is what actually happened.

What is certain is that there is a great deal of research into killer robots, that they are in the making, that there is much speculation, debate, and protest surrounding them. Also certain is that they connect to an ancient fantasy. For example, in the *Mahabharata*, a religious epic from the fourth century B.C., one can read that the enemies of the Hindu god Krishna sought help from the demons to make an air chariot with wings and iron sides. Those chariots ascended into the sky until the moment they sighted the followers of Krishna. There, they aimed their missiles at the followers and mercilessly killed them (Schwarz, 2018).

Killer robots are thus examples of autonomous AI systems, just as the self-driving cars of Uber and Google are. For other examples of such systems, think of Sony's 1999 mobile robot, specifically AIBO, a form of toy that can also act as a substitute for a pet in small town apartments. One of the reasons for this is that the robot is capable of learning. It learns to respond to phrases or learns that its programmed walking motion is not ideal for moving around the house. Some other examples: an AI system capable of detecting lung cancer on its own, which is very useful in underdeveloped areas where there are not enough radiologists; elevators in skyscrapers that try to reduce waiting time and actual travel time by analyzing and interpreting data on traffic flows, among other things; technology used by a bank that estimates your creditworthiness and determines whether or not you will get a loan; a platform such as Amazon Mechanical Turk that receives a message from a company looking for a worker for a small task, scours its membership database for the most suitable candidate, checks his or her work, and finally deposits the money in the worker's account (or not, if it turns out, for example, that the deadline was not met).

That these systems are autonomous means two things. First, you can interpret 'autonomy' in a negative way in this context. In that case, it means that these decision systems can function in a completely autonomous way, without any human intervention. They differ in that respect from most drones, which can do some things independently but are still controlled by human operators. Indeed, not only is it the case that humans are not needed, but often humans can no longer influence them once those systems are operationalized, for example, because human responsiveness is too slow or because the technology is too far away. Second, 'autonomy' also has a more positive meaning here. It also means that technologies can make decisions on their own, decisions that are not programmed by the designer, and thus can be unforeseen. Autonomous AI is clearly different from land mines in this respect. The latter technology does operate without human intervention, but according to a fixed schedule over which it itself has no control.

Nobody Is Responsible

The reason we refer to such AI has everything to do with the now famous text 'The responsibility gap' by philosopher Andreas Matthias (2004). In it, it is claimed that one cannot hold anyone responsible for errors caused by such systems. According to Matthias, if AIBO learned to walk on its own and collides with a child, seriously injuring it, you cannot hold anyone responsible for that. The same would apply to Tesla's self-driving car. No one is responsible for the victim of a collision between the self-driving car and a truck in 2016, according to Matthias (and Tesla). And suppose a killer robot is trained to kill a terrorist but ends up killing an innocent person, you would be unable to hold anyone accountable for that either. Autonomous AI systems, the claim goes, involve a responsibility gap: although you might want to hold someone responsible, you cannot—after all, there is no responsible person.

Why is that relevant to this chapter? Matthias believes that the responsibility gap is new. For all the decisions that were ever made, for all the actions that were ever done, and for all the things that were ever made, a responsible person could be designated who could be punished in case something undesirable takes place. Now, however, with the advent of AI, it is the first time in history that people—engineers, manufacturers, politicians—cannot be held responsible for the mistakes caused by their own products: things that make decisions and function independently.

Moreover, Matthias believes, this new reality is undesirable. The responsibility gap is not only novel but also an ethical problem. It is problematic that we cannot assign responsibility when deaths occur as a result of the use of autonomous cars and weapons, a walking toy robot or an autonomous elevator. See here the relevance of Matthias's reasoning: the problem he thinks that he detects, the seventh original sin, would be a new problem, one that turns the existing moral practice upside down and punches a hole in the realm of ethics. If this is true, then it is an argument in favor of the disruption thesis.

There are good reasons to ban killer robots—for example, they are not good at distinguishing civilians from combatants. There are also reasons to be at least cautious about autonomous decision systems in law or medicine, such as the risk of bias and unfairness. Still, we disagree with Matthias. There is no responsibility gap in the case of autonomous systems, and should there be one, it is not new.

That is what we argue later. The argument will be significantly longer than the previous sections because the theme of responsibility is well suited to illustrating how to develop moral reasoning, to showing what exactly virtuous reasoning means in the context of AI ethics. Moral responsibility is also a theme that crops up in the other chapters. Furthermore, it is also the case that responsibility is not only a complex topic but also a particularly relevant theme in the context of AI and far beyond, especially today. Finally, it is not always clear what is and is not meant by responsibility and the responsibility gap. We therefore first shed some light in this conceptual darkness.

What Is Responsibility?

To begin with, we would like to point out that responsibility is linked to numerous concepts: guilt, punishment, and so on. It is also important to see that 'responsibility' can be interpreted in several ways. When you say 'I am responsible' you can mean more than one thing. This is not a philosophical fabrication, but simply how everyday life works. It is, however, the philosopher who can draw attention to that multiplicity.

We focus on three meanings, which will come in handy later and to which we apply the following terms: causal responsibility, moral responsibility, and role responsibility. These terms already make it clear that we are not entering the domain of the law here, and therefore we are not talking about liability or legal responsibility. But what exactly do these three concepts mean?

Suppose a scientist works in a laboratory and uses a glass tube with toxic substances in it. When these substances are released, they spread throughout the building, resulting in the death of many colleagues. Normally, the scientist is careful, but a fly in the eye causes the scientist to stumble. The result is that the glass tube breaks and the toxins are released, causing deaths. Asked who is responsible for the havoc, some will answer that the scientist is. They then understand 'responsibility' in a well-defined sense, namely in a causal sense. They mean that the scientist is (causally) responsible because he or she plays a role in the course of events leading to the undesirable result.

Let us make a slight modification. The same scientist works in exactly the same context with exactly the same toxic substances, but now also belongs to a terrorist group. He or she wants the colleagues to die, and therefore deliberately drops the glass tube, resulting in several people dying. We again hold the scientist responsible, but the content of this responsibility is clearly different from the first kind of responsibility. Without the scientist's morally wrong act, the colleagues would still be alive, and so the scientist is the cause of the colleagues' deaths. So, he or she is certainly causally responsible. But to this form is added a second kind of responsibility. We say that the scientist is also morally responsible, and in this case, we mean that he or she should be punished.

Moral responsibility usually refers to one person, although it can also be about a group or organization. That person is then held responsible for something. Often this 'something' is undesirable, such as death, but you can also be held responsible for good things, such as saving people. If a person is morally responsible, it means that others can respond to that person in a certain way: praise or reward when it comes to desirable things; disapprove or punish when it comes to bad things. In addition, if one were to decide to punish or reward, it would also mean that it is right to punish or reward that person, that, in other words, there are good reasons to punish or reward that person, and not someone else—later, we go into those reasons in detail. Note that moral responsibility does not necessarily involve punishment or reward. It only means that someone is the rightful candidate for such a response, that punishment or reward may follow. So, one may be responsible for something undesirable, but what happened was not so bad that someone should be punished (Fischer & Ravizza, 1998).

The third form, role responsibility, refers to the duties that come with a role or position. Parents are responsible in this sense because they must ensure that their children grow up in a safe environment, just as it is the role responsibility of a teacher to ensure a safe learning environment. Or again, take the scientist from earlier. You can say that he or she is responsible, without referring to a role in a chain of events (causal responsibility) or to the practice of punishment and reward (moral responsibility). Those who believe that the scientist is responsible may thereby also refer to the duty to watch over the safety of the building: he or she must ensure that the room is properly sealed, that the glass tubes do not have cracks, and so on.

There are different relationships between the types of responsibilities. We look only at these that are relevant to the remainder of our story. The preceding paragraphs make it clear that a person can be causally responsible without being responsible in a moral sense. You do not condemn the scientist who trips over a shoelace. The reason is that he or she had no bad intentions. Conversely, though, moral responsibility always rests on causal responsibility. You cannot hold someone morally responsible if they are in no way part of the process that led to the (un)desired result. That causal involvement, by the way, should be interpreted in a broad sense. Suppose the scientist is following an order. The person who gave the order is then not only causally but also morally responsible, despite the fact that he or she did not commit the murder himself. Finally, role responsibility is always accompanied by moral responsibility. If, for example, as a scientist it is your duty to ensure that the laboratory is safe, it follows at least that you are a candidate for punishment if it turns out that you have not done your duty adequately, or that you can be rewarded if you have met expectations.

Tour of Duty

Autonomous systems lead to a responsibility gap, some claim. But what does one understand by responsibility here? Clearly, one is not talking about causal responsibility. AI systems are normally created by humans—we write 'normally' because there is already AI designing AI—and so it would express a problematic view of technology should you claim that no humans are involved in the creation of AI systems.

The responsibility gap is also not about the third form of responsibility, namely role responsibility. That too is obvious, and it has to do with the precautionary argument, which we cited in Chap. 1. That argument refers to the duty of engineers, not so much to create more sustainability or well-being, but to make things that have as little undesirable effect on moral values as possible, and thus to think about such possible effects in advance. As there is no reason why this should not apply to the developers of autonomous systems, the responsibility gap does not mean that AI systems have no special duties attached to them.

There is even reason to argue that that moral imperative applies primarily to the developers of AI systems. Because the decision-making power is being transferred to that technology, and because it is often impossible to predict exactly what decision will be made, they certainly must assume their responsibility, their role responsibility. That is, the developers of autonomous AI must think carefully, more carefully than other tech designers, about the possible undesirable effects that may result from the algorithms' decisions in the legal or medical world. With that explanation, we are not just resuscitating the central idea from *Das Prinzip Verantwortung* by philosopher Hans Jonas (1979), who wrote just about the first major work on the ethics of technology and who suggested that in a modern world, the effects of technology are so uncertain that designers need to think about the consequences even more so than before. With our explanation, we also turn the proposition around. Suppose it is true that the introduction of autonomous AI is accompanied by a responsibility gap, then at least it is not about role responsibility. On the contrary, such technology precisely affirms the importance of moral duties (Van de Poel & Royakkers, 2023).

So the thesis of someone like Matthias is about moral responsibility, and we can clarify it as follows: in the case of mistakes made by autonomous AI, there is no candidate for punishment, and there is no justification for making someone pay for it. You may have a spontaneous tendency to reward or punish someone, but that tendency has no suitable purpose, according to someone like Matthias. That is the thesis. But is it true?

All Hands on Deck!

Before we examine whether Matthias' thesis holds water, it is useful to briefly touch upon the difference between the alleged problem of the responsibility gap and three other problems. The first problem is reminiscent of a particular view of God, the second is the so-called problem of many hands, and the third is related to the idea of ethical or responsible AI.

It is spring and you are strolling through the city with your lover. You are enjoying a lovely sunny afternoon, until you step in chewing gum. You feel it immediately: with every step, your shoe sticks a little to the ground and it feels like your sole is no longer flat. Your mood changes, the sunny afternoon is gone (at least for a while) and you look to call someone on it. However, the person who left the gum on the ground is long gone. There is definitely someone causally responsible here: someone dropped the gum at some point. And the causally responsible person is also morally responsible. You are not supposed to do that, and if you do it anyway, then you are ignoring your civic duty and your being reprimanded is justified. However, the annoying thing about the situation is that it is not possible to detect the morally responsible person.

The problem in this example is that you do not know who the morally responsible person is, even though there is a responsible person. This is reminiscent of the relationship between man and God as described in the Old Testament. God created the world, but has subsequently distanced Himself so far from His creation that it is impossible for man to perceive Him. In the case of the responsibility gap the problem is of a different nature. There, it is not an epistemic problem, but an ontological problem. The difficulty is not that it is unknown who is responsible; the problem is that there is no one responsible for the errors caused by an autonomous system, so the lack of knowledge cannot be the problem here.

To illustrate the second problem that deviates from the responsibility gap, the problem of many hands, we turn to the disaster of the Herald of Free Enterprise, the boat that capsized on 6 March 1987, resulting in the deaths of nearly 200 people. An investigation revealed that water had flowed into the boat. As a result, the already unstable cargo began to shift to one side. This displacement eventually caused the ferry to disappear under the waves just outside the port of Zeebrugge in Belgium. This fatal outcome was not the result of just one cause. Several things led to the boat capsizing. Doors had been left open, the ship was not stable in the first place, the bulkheads that had been placed on the car deck were not watertight, there were no lights in the captain's cabin, and so on. Needless to say, this implies that several people were involved: the assistant boatman who had gone to sleep and left the doors open; the person who had not checked whether the doors were closed; and finally, the designers of the boat who had not fitted it with lights.

There are so many people involved in this case that not only one person can be held responsible. But that is not the same as saying that no one is responsible. The case is not an example of an ontological problem; there is no lack of moral responsibility in the case of the capsized ferry. Indeed, there are multiple individuals who are morally responsible. There is, however, an epistemic problem. Not that one does not know who is responsible and that it is impossible to find out, as in the example of the chewing gum. No, the problem is that there are so many hands involved that it is very difficult if not impossible to know exactly who is responsible for what and to what extent each person involved is responsible. Compare it to a string orchestra: because there are so many strings, it is almost impossible to pinpoint which string is responsible for which sound. Thus, in the case of the Herald of Free Enterprise, many knots had to be untangled in terms of moral responsibility, but that is different from claiming that the use of a technology is associated with a responsibility gap.

Finally, the claim that no one is responsible for the mistakes made by autonomous AI should not be confused with the claim that autonomous decision systems are ethically irresponsible AI systems. What exactly is the difference?

We pointed out earlier that in the context of ethical reflection on AI, different designations are used: Transparent AI and Ethical AI, but also Trustworthy AI and Responsible AI. The last three can be seen broadly as synonyms of each other. They refer to the use of AI that is morally unproblematic. Transparent AI, however, does not mean the same as ethical AI. Transparency is desirable for AI in many cases. If it is unclear what the reasons are for the decisions made by AI then the technology may face a moral problem, at least when that decision has some impact on people's lives. But transparency is not a sufficient condition for AI to be ethical. To be morally sound, those systems must also be neutral or nonbiased, sustainable, secure, and privacy proof. According to some, like Matthias, a necessary condition for Ethical AI, and therefore also for Trustworthy AI or Responsible AI, is that you must be able to hold someone responsible in case a mistake occurs. Moral responsibility would be a requirement for Ethical AI alongside neutrality and sustainability, among others. If there is a responsibility gap, then your system is ethically flawed.

This explanation makes it clear that there is a difference between irresponsible AI and the responsibility gap. The latter means that it is unjust to punish someone when autonomous AI makes a mistake. According to some, the consequence is that the technology is generally unethical, even if it turns out that it is neutral or transparent. The reason is that moral responsibility, though insufficient for AI ethics, is seen as necessary for ethics. But the reverse is not the case. It is not because a system is ethically irresponsible that there is a responsibility gap. This is because the attribution of responsibility, though necessary for responsible AI according to a number of thinkers, is not a sufficient condition for ethical AI. So, it is possible that there is someone morally responsible for an error caused by AI, but that the system is still not ethically responsible. This could be because the technology is biased, but also, for example, because it is not sustainable or secure.

Can Robots Suffer?

Artificial intelligence is a disruptive technology. That is the proposition at stake. More specifically, the question is whether there are reasons for this on an ethical level. Does autonomous AI lead to new moral problems, such as a responsibility gap, for example? If so, this could be an argument in favor of the disruption thesis. But is it true that there is no responsibility for mistakes made by autonomous AI?

There is an answer to that question that is often either not taken seriously or overlooked. It is about the possibility of AI systems being responsible themselves. To be clear, this is about moral responsibility and not the causal type. After all, autonomous technologies very often play a causal role in a chain of events with an (un)desirable outcome. Our question is: is it utter nonsense to see AI as the object of punishment and reward, praise and indignation?

One of the sub-domains of philosophy is philosophical anthropology. A central question in that domain is whether there are properties that separate humans from, say, plants and nonhuman animals, as well as from artificial entities. One can think in that context of the ability to play and communicate, to suffer psychologically, or to get gray hair. However, it is almost impossible not to talk about responsibility in that context. After all, we only attribute that to people today. Sure, we do not hold people with mental disabilities or disorders responsible for a number of things. And yes, we also punish and reward animals that are not humans. And we know that animal trials were held in the Middle Ages—dogs and pigs, among others, could be charged with crimes and subsequently buried or drowned. But moral responsibility today is something we reserve exclusively for humans, and thus do not attribute to artifacts. When the candy machine does not give you the bag of candy you paid for, you may feel anger, but normally you realize almost immediately that such a thing makes no sense. You do not hold a candy machine responsible, just as you do not attribute responsibility to a refrigerator—or to a tree or a dog.

Part of the reason we do not hold artificial entities responsible has to do with what responsibility is. We recall that a responsible person is the justifiable target of moral reactions such as punishment and reward, anger and indignation. Those reactions do not necessarily follow, but if they do follow then the responsible person is the one who is justifiably the subject of such a reaction. But that presupposes the possibility of some form of sensation, the ability to be affected in the broad sense, whether on a mental or a physical level. There is no point in designating someone as responsible if that person cannot be affected by the moral reactions of others. Of course, the ability to experience pain or pleasure in the broad sense of the word is not sufficient to be morally responsible. This is evident from our dealings with nonhuman animals: dogs can experience pain and pleasure, but we do not hold them responsible when they tip a vase with their tail. However, the ability to be physically or mentally affected by another's reaction is a necessary condition. And because artifacts such as autonomous technologies do not currently have that ability, it would be downright absurd to hold them responsible for what they do.

On the other hand, moral practices are not necessarily fixed forever. They can change over the course of history. Think about rights. At the end of the eighteenth century, people were still arguing against women's rights as follows: if we grant rights to women, then we must also grant rights to animals. The concealed assumption was that animal rights are unthinkable. Meanwhile, it is completely immoral to deny women rights that are equal to those of men. Take the robot Sophia. In October 2017, the unthinkable happened: Saudi Arabia granted Sophia citizenship. If throughout history more and more people have been granted rights, and if other moral practices have changed over time, why could there not be a change when it comes to moral responsibility? At the time of writing, we cannot hold artifacts responsible; but might it be possible in the future?

That question only makes sense if it is not excluded that robots in the future may be affected on a physical or mental level, that technologies may later experience pain or pleasure in some way. If that ability can never exist, then it is out of the question that our moral attitudes will change, then we will never hold AI morally responsible. And exactly that, some say, is the most realistic scenario: we will never praise technology because it will never be capable of sensation on a physical or mental level. Much can be said about that assertion. We only give a brief response to the following thought experiment that is sometimes given to support that claim (and that is based upon John Searl's famous thought experiment).

Imagine a humanoid robot, a robot that has the same shape as that of an average human body. It is impossible to tell the difference between a human and this robot based on physical appearance, as in the *Black Mirror* episode 'Be Right Back.' Moreover, you are in the head of that robot. You have to make sure that all the decisions, movements, and actions of the robot are indistinguishable from those of a human. In other words, you have to come up with the appropriate output for each incoming signal. If the robot is asked for a name, you have to give the signal that the robot identifies itself as, say, Sophia. Now suppose the robot falls down the stairs, a situation that in humans would normally be accompanied by the sensation of pain. You provide the output that usually follows pain: yelling, crying, and so on. The robot looks like a human, falls, and then reacts as humans normally do. However, is the robot in pain? Someone may react to the fall, for example, because it is a reflex to react to signs of pain. However, no one will react because the robot is in pain. Although the robot does show signs of pain, there is no pain, just as computer programs such as Google Translate and DeepL do not really understand the Chinese sentence that they can nevertheless translate perfectly.

Smart software programs should be compared to the human in the robot's head: they provide the output, the pain signals. First, they interpret numerous data and then when they receive input, they predict the best outcome based on the learning phase. Applied to this thought experiment: an intelligent robot that first analyzes photos and videos of people in pain will, in the event of a fall, produce the signals that normally express pain in humans. Can we then conclude on the basis of the thought experiment that AI systems will never be able to experience pain?

It is worth distinguishing what the thought experiment does and does *not* tell us. On the one hand, it suggests that pain might not be the same as a pain signal, that having pain is not identical to expressing pain. You can have pain without expressing it, and you can produce pain signals while not actually being in pain. In other words, the experiment brings to the surface that you cannot infer from the fact that smart technology produces pain signals that the technology is actually in pain. AI can produce things that indicate pain in humans, but those signals, in the case of the software, are not in themselves a sufficient reason to conclude that the technology is in pain. This is what we can gather from the case with the falling robot.

On the other hand, the thought experiment does not show that technologies will never be able to feel pain, let alone prove that machines will never be able to be affected mentally. Certainly, it is quite possible that robots will never be able to feel pain or anything, but that is not what the thought experiment makes clear to us. Furthermore, the experiment is only relevant to software, whereas technologies usually consist of hardware as well. And this hardware is precisely the reason not to immediately cast aside the possibility of pain. At least, that is what emerges from yet another thought experiment.

Like all physiological systems of a human body, the nervous system is made up of cells, mainly neurons, which constantly interact. This causal link ensures that incoming signals lead to the sensation of pain. Suppose, however, that you are beaten up, and that for 60 min you are actually in pain—whether you express that pain in physical or behavioral signals is not important here. Suppose, moreover, that science has advanced to the point where the neurons can be replaced by a prosthesis, microchips for example, without it making any difference otherwise. The shape of the prosthesis is different and the material is also different—the chips are made on a slice of silicon—but otherwise those artificial entities do exactly the same thing as the neurons: they send signals to other cells and provide sensation. So, you have pain, you have the body of an average human being, and you will continue to feel pain for the next hour. Only for 60 min, a scientist will replace every cell with a microchip, so that your body is not only made up of cells but also of chips. In other words, you float like a cyborg somewhere between natural and artificial, a bit like Arnold Schwarzenegger in *The Terminator*. Is it still utter nonsense to claim that robots might one day be able to feel pain?

The foregoing is a thought experiment, but it is by no means complete nonsense. Research has been set up in the past to induce emotions in robots. To avoid confusion, we would like to stress the following: we are not claiming that intelligent systems will one day be able to feel pain, that robots will one day resemble us—us, humans—in terms of sensation. At most, the last thought experiment was meant to indicate that it is perhaps a bit short sighted to simply brush this option aside as nonsense. Furthermore, we are not claiming, if it does turn out that AI systems can experience pain, that we will automatically hold them responsible for the things that they do. The reason is that the ability to feel pain is not enough to be held responsible. Our relationships with nonhuman animals, for example, demonstrate this, as we pointed out earlier. Suppose, however, that all conditions are met, would that immediately imply that we will see autonomous AI systems as candidates for punishment and reward? In any case, attributing responsibility exclusively to humans is an age-old moral practice, which is why this may not change any time soon. The latter is something one might expect on the basis of the history of moral rights. That history may show that moral practices are not necessarily eternal, and also that the time-honored practice of attributing rights only to humans is only gradually changing in favor of animals that are not humans. That alone is a reason to suspect that ascribing moral responsibility to robots may not be for tomorrow, even if robots could be touched physically or mentally by reward or punishment.

Child Soldiers and Drones

So we must return to the central question: do autonomous AI systems create a responsibility gap? If so, then we have an argument for the disruption thesis. At least that is what someone like Matthias claims, because such a gap does not exist anywhere else, and moreover, such a thing would be undesirable. But do smart systems create such a gap? Technologies themselves cannot be held responsible today, but does the same apply to people?

There is reason to suspect that you can hold people responsible for mistakes made by autonomous AI. Consider an army officer who engages a child soldier. The child is given a weapon to fight the enemy with. But in the end, the child kills innocent civilians, thus committing a war crime. Perhaps no one would say that the child is responsible for the civilian casualties, but in all likelihood we would believe that at least someone is responsible. Normally, we hold the officer morally responsible for the crime. However, is there a difference between this case and the use of autonomous AI systems, killer robots for example? If so, is that difference relevant? Of course, child soldiers are human beings, robots are not. In both cases, however, a person makes a decision knowing that undesirable situations may follow and that one can no longer control them. If the officer is morally responsible, why should the same not apply to those who decide to use autonomous AI systems? Are killer robots and other autonomous AI systems something exceptional in that regard?

Should it turn out on the other hand that someone like Matthias is right, that there is indeed a responsibility gap, that too would probably not be entirely surprising. This is partly because moral responsibility has a gradual character: it can increase or decrease. Consider, for example, the trainer of a sports team. It goes without saying that he or she is at least partly responsible for the team's performance. However, does that also mean that the degree of responsibility is always the same? No, the coach can be held responsible to a lesser extent than last season for the poor performance of his or her team, for example, because the board allowed some good players to leave and instead bought players who have fewer qualities. What does that have to do with the so-called responsibility gap?

Let us make this more concrete and focus on war technology. Suppose you are a soldier and kill a terror suspect. If you used a classic weapon that functions as it should, a 9-mm pistol for example, then without a doubt you are entirely—or at least to a large extent—responsible for the death of the suspect. Suppose, however, that you want to kill the same person, and you only have a semiautomatic drone. You are in a room far away from the war zone where the suspect is, and you give the drone all the information about the person you are looking for. The drone is able to scout the area itself, and when the technology indicates that the search process is over, you can assess the result of the search and then decide whether or not the drone should fire. Based on the information you have gathered, you give the order to fire. But what actually happens? The person killed is not the terror suspect and was therefore killed by mistake. That mistake has everything to do with a manufacturing error, which led to a defect in the drone's operation. Of course, that does not imply that you are in no way morally responsible for the death. So there is no responsibility gap, but probably most people would feel justified in saying that you are less responsible than if you used a 9-mm pistol. This has to do with the fact that the decision to fire is based on information that comes not from yourself but from the drone, information that incidentally happens to be incorrect.

Thus, for many, the decrease in the soldier's causal role through technology is accompanied by a decrease in responsibility. The siphoning off of an activity—the acquisition of information—implies, not that humans are not responsible, but that they are responsible to a lesser degree. This fuels the suspicion that devolving all decisions onto AI systems leads to the so-called responsibility gap. But is that suspicion correct? If not, why? These questions bring us to the heart of the analysis of the issue of moral responsibility and AI.

Conditions of Responsibility

As previously announced, we want to take a conservative position. Although autonomous AI systems are new, the differences from classical non-autonomous technologies are not such that for mistakes made with the former we cannot hold anyone responsible, and for mistakes made with the latter we can hold someone responsible. The existing moral practice is not interrupted by new technologies; there is no moral disruption in terms of responsibility either. We suspect that many people agree with this but at the same time find it difficult to explain in detail why exactly they agree, why at least someone is responsible for the mistakes made by an autonomous system. Philosophy can help with this. To be brief: at least someone is responsible, and that is because the classical conditions for responsibility are also now fulfilled. We pointed earlier to the capacity for sensation in the broad sense of the word, but what other conditions must be fulfilled for someone to be held responsible? We distinguish causal responsibility, autonomy, and knowledge.

It goes without saying—we touched on this a few pages ago—that moral responsibility presupposes causal responsibility. Someone who is not involved at all in the creation of the (undesirable) result of an action cannot be held responsible for that result. In the context of AI, several people meet this condition: the programmer, the manufacturer, the user. However, this does not mean that we have undermined the responsibility gap theorem. Not every causal relationship is relevant to moral responsibility, not every involvement is associated with moral responsibility. Consider the scientist in the laboratory we discussed earlier. If she or he falls and as a result causes the death of a few colleagues, we do not hold that scientist responsible in a moral sense, only in a causal sense.

Thus, more is needed. Moral responsibility also requires autonomy. One can understand this concept in at least two ways. When someone talks about autonomy, it can be understood in a negative sense. The one who is autonomous in that respect can function completely independently. For our reasoning, only the second, positive form is relevant. This variant means that you can weigh things against each other, and that you can make your own decision based on that.

However, the fact that you are able to deliberate and decide is not sufficient to be held morally responsible. For example, you may make the justifiable decision to kill the king, but when the king is killed, you are not necessarily responsible for it, for example, because someone else does it just before you pull the trigger and independently of your decision. You are only responsible if your decision is causally linked to the murder; the deliberate decision must be at the root of the murder. In other words, when someone is morally responsible, it presupposes a causal link between autonomy and the act. It is only through that link that there is involvement in the act and that autonomy is relevant to moral responsibility.

Knowledge is the final condition. You can only be held responsible if you have the necessary relevant knowledge. One who does not know that an action is wrong cannot be responsible for it. And if the consequences of an act are unforeseen, then you cannot be punished for that either. Note that the absence of knowledge does not necessarily exonerate you. Indeed, you may not know certain things whereas you really should have known them. Formulated in terms of responsibility: if it is your role responsibility to know about certain things, but you do not, and that lack of knowledge leads to undesirable results, then you are morally responsible for the undesirable result. To take a simple example, if the driver of a car runs a red light and causes an accident as a result, then that driver is morally responsible for that accident, even if it turns out that he or she was unaware of the prohibition against running a red light. After all, it is your duty as a citizen and car driver—read: your role responsibility—to be aware of that rule.

Everything Under Control

So, whoever is involved in the use of a technology, whoever makes the well-considered decision to use that technology, and whoever is aware of the necessary relevant consequences of that technology: they can all be held morally responsible for everything that goes wrong with the technology. At least that is what classical analysis implies.

Matthias, however—and this is crucial—believes that an additional condition must be met. Once an action or certain course of events has been set in motion, he believes that you must have control over it. So even if you are involved, for example, because you have made the decision that the action or course of events should take place, although you can do nothing else about it at the time at which it was initiated, it would be unfair to punish you when it all results in an undesirable outcome. As AI systems function completely independently, in such a way that you cannot influence their decisions, then according to someone like Matthias, you cannot hold anyone responsible for the consequences.

In parentheses: it is important to keep in mind that this reasoning assumes that each person involved has taken on his or her role responsibility. Suppose you are a designer. You must therefore think about the possible negative consequences of the design and anticipate to as far an extent as possible. If you do not do so, then in other words, you are not fulfilling your role responsibility, and precisely that leads to the system causing a failure. In that case, it goes without saying that you can be held responsible for that undesirable consequence. That moral responsibility follows directly from the lack of role responsibility, even when you no longer have control. Matthias also agrees.

There are things that seem to support Matthias' argument. If you are held responsible for an action, it usually means that you have control. As CEO, one is responsible for the company's poor numbers because one could have made different decisions that benefited the company more. Conversely, if one has no control over a large number of factors, then this usually implies that one bears no responsibility for them. For example, you have no control over the weather conditions, nor do you bear any responsibility for the consequences of good or bad weather. Thus, responsibility is often accompanied by control, just as the absence of control is usually accompanied by the absence of responsibility. Yet it is false to say that you *must* have control over an initiated action or course of events to be held responsible, and that not having control takes away your responsibility. Although it is often said that today we want control over just about everything and that control is very important to us moderns, our moral practice tells us otherwise. Control is not so important today that it removes our moral responsibility if we do not have control over things. This is demonstrated by the following.

Imagine you live in the center of London and you want to spend an evening with some friends at a café 10 km away in North London. It is raining. Not wanting to get wet, you decide to take the car. You set off, but after a few minutes you suffer from an epileptic fit. This is not the first time, because you are an epileptic. The result of the seizure is that you lose control of the steering wheel, causing the car to go off the road and eventually seriously injuring a cyclist. It is not certain that you will be punished, let alone receive a severe sentence, but perhaps few, if any, will not hold you responsible for the cyclist's injury, in spite of your lack of control of the car's steering wheel. Why?

First of all, you possess all the relevant knowledge. You do not know that a seizure will occur within a few minutes, but as someone with epilepsy you do know that there is a risk of a seizure and that such a thing may be accompanied by an accident. Furthermore, you are autonomous (in a positive sense). You are able to weigh up a number of things—the desire to go to a café, the desire to not get wet, the risk of an attack—and to make a decision based on that. Finally, you purposefully get in the car. As a result, you are causally connected to the undesirable consequence in a way that sufficiently grounds moral responsibility. After all, if you make a decision knowing that it may lead to undesirable consequences, then you are justified in considering yourself a candidate for punishment at the time the undesirable consequence actually occurs. Again, it is not certain that punishment will follow, but those who take a risk are responsible for that risk, and thus can be punished when it turns out that the undesirable consequence actually occurs.

It is important for the sake of our argument to know what one can and cannot infer from this. The only thing you can conclude from the above is that not having control does not absorb your moral responsibility. Responsibility does not require control, and so you cannot say that autonomous AI is associated with a responsibility gap because you have no control over the technology. From the foregoing, however, you cannot conclude that the idea of a responsibility gap in the case of autonomous AI is incorrect, that in all cases someone is responsible for the errors caused by that technology. After all, perhaps in the case of autonomous decision systems the other conditions for moral responsibility are not met and we should still conclude that the use of autonomous AI goes hand in hand with a responsibility gap, even if it turns out that control is not required for moral responsibility. But is this the case?

Breaking Pot, Paying Pot

In addition to our critique of Matthias, we want to make another point. We want to show that you have to understand the use of this hyper-advanced technology in the same way as the situation with the epilepsy patient just now. There is at least someone responsible when autonomous AI makes mistakes—maybe there is even collective responsibility—but because it is enough to find one person responsible to undermine the thesis of a responsibility gap, we do not need to go into the idea of shared responsibility. To show that it is not true that no one is responsible, we invoke some previously cited examples: a killer robot kills a citizen, and a self-driving car hits a cyclist.

To begin with, it is important to note that both dramatic accidents are the result of a long chain of events that stretches from the demand for production, through the search for funding, and finally to the programming. If we are looking for a culprit, we might be able to identify several people—we are thinking of the designer or producer, for example—but the most obvious culprit is the user: the commander who decides to deploy the killer robot during a conflict, or the occupant of the autonomous car. It is justified to put them forward as candidates for punishment for the following reasons, just as the epilepsy patient is responsible for the cyclist's injury.

First of all, both know the context of use and what the possible undesirable consequences could be. They do not know whether or not an accident will happen, let alone where and when exactly. After all, autonomous cars and weapons are based on machine learning and neural networks, which means that it is not (always) possible to predict what decision will be made. But the kinds of accidents that can happen are not unlimited. Killing civilians and destroying their homes (killer robot), hitting a cyclist or crashing into a group of people (self-driving car) are dramatic but foreseeable; as a user, you know such things can happen. And if you do not know, that is a failure on your part: you should know. It is your duty, your role responsibility, to consider the possible negative consequences of the things you use.

Second, both commander and owner are sufficiently autonomous. They are able to weigh up the advantages and disadvantages: the chance of war crimes and fewer deaths in their own ranks (killer robot), the chance of traffic casualties and being able to work while on the move (self-driving car).

Third, if, based on these issues, the decision is made to effectively move to the use of autonomous cars and weapons, though knowing that it may bring undesirable consequences, then it is justifiable to hold both the commander and owner responsible for deliberately allowing the undesirable anticipated consequences to occur. Those who take risks accept responsibility for that risk; they accept that they may be penalized in the event that the unwanted, unforeseen consequence actually occurs (Hansson, 2013).

Thus, in terms of responsibility, the use of autonomous AI is consistent with an existing moral practice. Just as you can hold people responsible for using non-autonomous technologies (or for the damage caused by their pets), people are also responsible for things over which they have no control but with which they are connected in a relevant way. So not only does the autonomy of technology not erase the role responsibility of the user it does not solve moral responsibility either. In short, and with this we pick up the thread of this chapter again, even when it comes to moral responsibility, AI systems do not provide an argument in favor of the disruption thesis. The path that the system takes to make a decision may be completely opaque to the user, but the system does not create a responsibility gap and thus does not create a new moral problem.

Those who disagree must either demonstrate what is wrong with the existing moral practice in which we ascribe responsibility to people or demonstrate the difference between moral responsibility in the case of autonomous systems and everyday moral practice. By the latter we mean the relevant differences. For of course there are differences between using an autonomous system on the one hand and driving a car as a patient on the other. The question, however, is whether those differences matter when it comes to moral responsibility.

Acting Under Coercion

So, when it comes to AI, there are not seven but six deadly sins. And like bias and privacy, among others, moral responsibility is not an argument that supports the disruption thesis. This is true, however, only insofar as our analysis in the past few pages makes sense. But assume for a moment, to conclude this chapter, that what we have just argued is not true: there is a responsibility gap. Suppose that the analysis we gave earlier goes wrong in several places, and that you really cannot hold anyone responsible for the damage caused by the toy robot AIBO, Google's self-driving car, Amazon's recruitment system, or the US army's killer robot. In that case, would that make an argument for the disruption thesis? In other words, would a responsibility gap in the case of autonomous AI be unique?

If the responsibility gap exists in the case of AI, it does not exist only in that context. When lightning strikes a tree that then falls on your car, it involves damage, but you cannot hold anyone responsible for it. Does the gap also exist when people themselves are causally involved in an undesirable event? Suppose you are working as a bank teller, and someone overpowers you with a gun. With the gun to your head, you are forced to open the safe. Even though you do not want to, you fear for your life and open the safe for the robber who walks away with a lot of money. Because you are forced with a gun to your head, no one will hold you responsible. But there is no responsibility gap: the robber must be punished.

Let us make a change. You are still the bank teller; you open the safe again and you make a lot of money disappear. And although you are not being coerced by anyone else, even now you are not acting autonomously. You are delusional for the first time. There is a voice compelling you to open the safe and run off with the money. Thus, you are causally responsible for the robbery, but no one is morally responsible for it. At least as far as no one knows all the facts; if one knows what mental state you were in, your responsibility disappears. Moreover, you robbed the bank at night, when no one was there. And even after that, when you hid the money, you acted entirely alone. There is no one who helped you even a little. Conclusion? There is an undesirable situation for which you want to but cannot designate a responsible person, not because you do not know who the responsible person is, but because there is no responsible person. It would be completely unjustified to punish anyone for what has happened. In other words, should AI be accompanied by a responsibility gap, such a gap would not be novel.

Conclusion

Perhaps there are quite a few people—we are thinking mainly of entrepreneurs and technologists—who defend the thesis that AI is a disruptive technology because they have vested interests in it, especially economic interests. Designers and companies want to bring a new technology to market, and in order to convince potential users and buyers, one claims that the new design will turn an existing domain upside down or disrupt it. If this hypothesis is correct, then it is clear that we are dealing with a particular interpretation of the disruption thesis, one concerning the positive disruptive effects of AI. The interpretation we focused on in this chapter has a rather negative interpretation. We zoomed in on the problems caused by smart technology, more specifically its moral problems. The interpretation is that AI causes entirely new moral problems, problems that were not there when AI did not exist. Such an assertion, of course, is not interesting for techno-optimists, but rather fits the worldview of alarmists and techno-pessimists. We do not claim that no new moral problems will arise in the future, but the ethical problems that AI faces today are certainly not new. Privacy, bias, and transparency, for example, are the issues most often talked about today when it comes to AI ethics, but problems with privacy, bias, and transparency are not solely confined to AI. In short, AI is not a disruptive technology, at least as far as ethics is concerned. This is the thesis that we have put forward and defended in this chapter.

Chapter 3
Technological Determinism

> *We shape our buildings, and afterwards, our buildings shape us.*
>
> Winston Churchill, 28 October 1943

In an interview for *Playboy* in 1969, media scientist Marshall McLuhan, known for, among other things, the expression that the medium is the message, claimed that computers will be able to orchestrate the lives of many people in the near future. Machines, the Canadian thinker said, will be able to take over the main media channels, write messages themselves, and distribute them to the population. Meanwhile, we know that McLuhan's prediction was not complete nonsense—think of Microsoft's chatbot TAY (the acronym stands for Thinking About You) that sent sexist and racist messages into the world via Twitter in March 2006. At the time, however, interviewer Eric Norden was somewhat impressed by what McLuhan was claiming. He asked if we as humanity will still have control over technology. Or, Norden queried, will technology control humanity? McLuhan replied as follows: "I see no possibility of a worldwide Luddite rebellion that will smash all machinery to bits, so we might as well sit back and see what is happening and what will happen to us in a cybernetic world. Resenting a new technology will not halt its progress." (Norden, 1969).

McLuhan, like many commentators on technology and artificial intelligence (AI), did not shy away from bold statements and media appearances, including a cameo in Woody Allen's 1977 film *Annie Hall*. His remarks in the now-famous interview with *Playboy* are particularly insightful. McLuhan argued that nothing, not even the neo-Luddites, could halt the relentless advance of technology. This reflects the determinism thesis, which, alongside the neutrality and disruption theses, is quite popular among engineers and scientists. The term 'Luddite' originates from Ned Ludd, a figure from the late eighteenth century rumored to have destroyed weaving machines in protest against job losses due to automation. In this chapter, we scrutinize the determinism thesis, which views technology and its development as necessary. Is technology inevitable?

To avoid misunderstandings, it is worth noting that there are at least four interpretations of the determinism thesis. McLuhan's interpretation is one of them. It reads, as one can gather from the quotation above, that the development of technology is inevitable. Once a technology is put on the market, it cannot but result in the production of new technology. The second interpretation is akin to the first and holds the following: all technology is necessarily created; that a technology is invented is inevitable. You can find this interpretation, as well as the first, for example, in *What Technology Wants* by Kevin Kelly (2011), the former editor-in-chief of the technology magazine *Wired*. The third interpretation of the determinism thesis is not so much about things we call 'technology,' but about a technological, instrumental way of looking at reality. It states that this way of thinking has taken over all domains of society, to such an extent that today we cannot look at things in any other way than the instrumental way. Perhaps the best known representative of this interpretation is the philosopher we already mentioned in Chap. 1: Heidegger. In his well-known 1954 text *Die Frage nach der Technik*, he claims that "[...] technology would be the fate of our time, where fate means the inescapability of an unchangeable outcome." (Heidegger, 1977). Finally, the last interpretation is about the connection between technology and society. Not everything we do today is determined by devices or the platforms of Google and Amazon, but technology does determine interpersonal relationships and social processes. One can find this assertion in Robert L. Heilbroner's famous essay 'Do Machines Make History?': "I think we can indeed say that the technology of a society imposes a certain pattern of social relations on that society" (Heilbroner, 1994).

From Printing Press to Protestantism

Later, we will explain these four interpretations of the determinism thesis in more detail. In the remainder of the introduction, we dwell only on the last one, which deals with technology and its social effects. That should suffice as a warm up.

The interpretation of the determinism thesis on social effects is reminiscent, among other things, of Plato's *Republic*. In it we read about Glauco recounting the myth of Gyges. Gyges is a good shepherd who finds a golden ring and notices that it has magical properties: when you turn the seal of the ring to the palm of your hand, you become invisible. The shepherd uses that ring to penetrate the court, seduce the queen, kill the king, and finally ascend to the throne. Glauco tells the myth to make clear that even a righteous person will eventually flout moral prohibitions when there is an advantage to be gained. There is no doubt that people with depraved souls will use the ring to secure their interests, but even gentle souls like the good shepherd Gyges will not fail to do so. In short, the ability to use an artifact (the ring) will lead to the same result in all cases. The fact that one can turn the ring to the palm of one's hand will always result in a moral mess, independent of human character, or more broadly, independent of culture.

Yet this version of the determinism thesis only became popular near the end of the nineteenth and the beginning of the twentieth century, under the influence of economist and sociologist Thorstein Veblen. The thesis that technology determines society surfaced in various contexts: for example, in advertising. In April 1920, an advertisement in the *Ladies Home Journal* featured a picture of the newest iron (the Simplex Ironer) next to the old version, and this with the message that the new technology will make a whole group of women happier (Smith, 1994).

The notion of technology determining society also returns, rightly or wrongly, when thinking about feudal society. Some probably argue that the introduction of gunpowder from China ushered in the decline of the social and political regime of the Middle Ages. This would have been due in part to the fact that knights on horseback were an easy target for soldiers armed with pistols. Because of that risk, knights used the horses and stirrups that they could borrow from peasants in exchange for protection less and less. Weapon technology, it is argued, would thus have directly and inevitably contributed to undermining the system of feudal lords and feudal peasants.

Another and often cited example of the determinism thesis is the claim that the printing press in the fifteenth century was bound to lead to the Reformation (Smith, 1994). In the period preceding that invention, the Bible was generally accessible only to the clergy. But through the famous innovation of Johannes Gutenberg, more and more people had access to God's word, and that, according to some, was bound to lead to the flowering of Protestantism and the schism in Christianity.

At the end of the nineteenth century, people were quite suspicious of the first telephone. It was thought that this technology would cause people to visit each other physically less often, so social contacts would undoubtedly become more superficial. Families would be scattered, people would only talk to like-minded people, do their work electronically, and only meet each other on ceremonial occasions, so it was said in 1893 (we note that this response is reminiscent of how some reacted to the rise of social media in the early part of this century).

Finally, the determinism thesis about the social effects of technology is regularly defended or cited in philosophical texts. For example, in his 1847 text *Misère de la philosophie*, Karl Marx writes the following: "The hand mill gives you society with the feudal lord, the steam engine society with the industrial capitalist." (cited in Heilbroner, 1994). Another illustration can be found in *Le système technicien* by Jacques Ellul. In that famous 1977 study, you can read the following: "Technology is autonomous from economics and politics. Technique provokes and conditions social, political and economic changes. Technology is the engine of everything else, despite any appearance to the contrary and despite human pride, which pretends that man's philosophical theories are still determining influences and man's political regimes are still determining factors in technical evolution." (cited in Pitt, 1987).

The Technological Condition

At first glance, there seems to be something to be said for the determinism thesis about the social effects of technology. There are hardly any places where you will not find clocks, for instance. Waking up from sleep, getting into a car, opening a laptop, looking at a church: it seems you cannot escape digital or analog clocks. And more generally, we use technology to eat (microwave), move (bike), sleep (Bose Sleepbuds II), study (laptop), climax (vibrator), write (pen), communicate (phone), advertise (social media), travel (plane), or heal (MRI scan). The cyborg is an amplification of what characterizes all or most human lives, namely that everyone's life is intertwined with technology through and through. The human condition is a technological condition. Is that not reason enough to suspect that technology inevitably has social effects?

Yet there are also things that tease the opposite. We think of the first computer, the Colossus. It was designed during World War II by engineer Tommy Flowers, who relied on the work of mathematician and computer scientist Alan Turing. This was in a military context. The goal was to be able to decipher the secret codes of the German army. During the Middle Ages, in monasteries, people should not pray to God separately but together, even during the night. Because people do not spontaneously wake up at the same time, this created the need for an alarm clock mechanism. The first mechanical clock is said to have been made around the tenth century by monk Gerbert of Aurillac, who later rose to fame as Pope Sylvester II. So in addition to the computer, the alarm clock also indicates that there is something wrong with the determinism thesis. Both show that technology arises under the influence of a social context, which is either religious or military in nature. They suggest that it is not technology that determines society, but the reverse, that society determines technology.

Two other examples: the bicycle and space travel. In his often cited study *Of Bicycles, Bakelites, and Bulbs: Toward a Theory of Sociotechnical Change from 1995*, technology sociologist Wiebe Bijker (1995) shows that in the beginning there were various types of bicycles and that this diversity was related to the difference in taste and preferences of people. The so-called Hoge Bi, a bicycle with a large front wheel, was especially popular with young men, because it enabled them to reach high speeds and could be used to impress women. Women found it dangerous and even immoral; a lower model with equal wheels was designed for them. Finally, there is also reason from space to doubt the version of determinism regarding social effects. Toward the end of the 1980s, funding for the development of technologies to explore space decreased. That had a lot to do with the end of the Cold War, with the decrease in nationalistic tension between the USA and the USSR.

With this introduction, we already have an idea of the debate around the determinism thesis, at least of one of the four interpretations. Do machines determine society? Does AI determine society? Do programmers run the world? If you include the other three interpretations of the determinism thesis, it becomes clear that the following questions will also be at issue soon. Is it true that a technological instrumental view dominates our culture? And if so, is this really as problematic as is often assumed? In addition, the following questions will also be addressed. Do technologies inevitably arise? Does one technology necessarily lead to another? Or is technology rather a response to a demand coming from society? These are the questions around which this chapter revolves, a chapter that is again hung on mainly two coat hooks, supplemented by a third.

First of all, we need to clarify a few things in the following pages. For example, what does 'determine' mean, and what exactly is to be understood by the determinism thesis? We clarify what the four interpretations do and do not mean. Without seeing this multifaceted nature, it is impossible or at least very difficult to properly understand and engage in the debate about the relationship between technology and society. Second, we will also evaluate. We argue that none of the four interpretations is correct. Not all technology arises necessarily; it is clear that not every technological evolution is inevitable; an instrumental view is not inevitable, nor is it certain that it is always problematic; and if technology has social effects, they do not necessarily always follow. Finally, we show why we think we need to discuss the questions that we raise. For example, if it is true that technology does not determine society but is itself determined by social factors, then that implies that the world could have looked different. Even if it remains to be seen whether that is a reason for much optimism, you can see from this that this chapter returns to a theme that is much older than thinking about technology, namely

that of necessity and contingency. Is everything that exists now inevitably so? Or can it be otherwise?

Determinism Determined

As indicated, there are four interpretations of the determinism thesis, four claims in which technology or things related to it are described in terms of determinism. But what exactly does 'determinism' mean? That is not always clear. This may be partly because, like 'justice' or 'responsibility,' the term can mean different things; and in the context of a philosophical reflection on technology, 'determine' has a specific interpretation. So we would do well to first explain what that term does and does not mean here.

Often 'defining' is given as a synonym for 'determining.' But it does not help us much yet, because what exactly is meant by that? The first obvious meaning of 'to determine' is 'to identify.' It is in this sense that biologists, for example, use the term when talking about a plant. They mean by 'to determine' to find out to which species that plant belongs. The determinism thesis, however, has nothing to do with this interpretation but with the second. When you use the other meaning, by 'to determine' you refer to a relation between two things, more precisely to a causal relation. It means that something, a process for example, has an effect on something else, or that something new comes into being as a result of that process. This interpretation is used when people say that lifestyle is a determinant of health. The way in which you live determines your health: it has an effect on your health, an effect that may or may not be positive.

However, the second meaning of 'to determine' does not fully coincide with 'to cause.' This is shown, for example, by the following. When you say that your interest in poetry is a result of the fact that your parents used to read a lot of poetry at home, that does not mean the same thing as saying that your interest is determined by your parents' interest. 'To determine' does deal with a relationship of cause and effect, but that description is not sufficient; a more precise, stronger description is required. This goes as follows. 'Determinism' refers to a particular type of effect that assumes a determinant, namely a necessary effect. It means that a state inevitably leads to a certain effect; that effect must follow from that state; the cause guarantees the effect. Even if we were able to rewind time, we would see that exactly the same initial state would produce exactly the same effect. The relationship between boiling and evaporation, for example, has such a necessary character. It is not merely that evaporation is an effect of boiling; the boiling of water cannot but result in evaporation. Or take Boyle's gas law: when the volume of air in a bicycle pump decreases, the air pressure inevitably increases.

A few centuries ago, mathematician Pierre-Simon Laplace still thought that everything is determined, that all causal relationships are necessary, that all states are an inevitable consequence of previous states. Now there is little or no belief in this. Some causal relations are patently necessary, but some are clearly not; certain effects might as well not be there. Take, for example, the influence of parents on children in terms of music choice. Quite a few people's interest in the Beatles follows from the fact that their parents used to listen to the Beatles a lot. Yet that effect is not inevitable. The interest might not have been there, even if the parents used to put 1969's *Abbey Road* on the turntable every weekend.

It is not important here whether they are all equally plausible or whether they make any sense at all, but to conclude we would like to point out that there are at least four forms of determinism, each with a different cause: a genetic variant, environmental determinism, psychological determinism, and a technological version. Only the latter will demand our attention, as will soon become apparent. After all, the four versions of the determinism thesis that we just mentioned in the introduction all have to do with the technological version of determinism (Pitt, 1999).

Genetic determinism means that a gene encoding a trait guarantees that an organism will exhibit that trait. Scientist Walter Gilbert defended this theory. He once said that we could predict a person's phenotype simply by analyzing their DNA on a CD-ROM. Today, hardly any scientists believe this. At best, genes make an effect probable, but they do not guarantee it. In addition to genetic determinism, there is an environmental determinism: one can determine the development of people on the basis of external influences. John Watson, the founder of behaviorism, is situated in this context. In 1926, he wrote that if a parent wishes his or her child to become a doctor, they can make sure that he or she becomes a doctor, and that is independent of the child's talents, ancestry, or desires. One form of psychological determinism is found in popularized introductions to Freudian psychoanalysis. There, you read that the child's conscious or unconscious rejection of the so-called Oedipus complex results in sexual perversion. If you do not accept the incest prohibition, the theory goes, then you become a fetishist or masochist, among other things. Finally, and most importantly, the determinism thesis is about the technological variant. In other words, it is about a relation in which an effect is inevitably produced, and about the place that technology, or things closely related to it, occupy in that relation. As already touched upon, the four interpretations of the determinism thesis fall under technological determinism. Not only is the claim that technology necessarily has this or that social effect an example of this but also the theory that deals with the instrumental thinking that characterizes our time. It is with the latter version that we begin.

Colonizing the World

If the determinism thesis can be understood in several ways, then this implies that the arguments used in this context are always an argument for or against a well-defined interpretation of the determinism thesis. It is important to keep this in mind to avoid confusion. After all, one person could argue for the determinism thesis and one person against it, but each on the basis of a different interpretation of that thesis, making it a dialogue of the deaf.

As already indicated, there are four interpretations of the determinism thesis. One deals with the social effects of technology, and two others with the developmental history of technology, more specifically the emergence and evolution of technology. We clarify and evaluate these in more detail later in this chapter. We begin with the interpretation of the determinism thesis that is in line with a popular, widespread view of our culture. This holds that today, we regard everything as a means of realizing our goals and satisfying our desires.

This brief description immediately shows why this version of technological determinism is an outlier. After all, it is not concerned primarily with technologies, things with a function, but with a way of looking and thinking. Yet, that theory comes under the heading of technological determinism because it is about a technological way of looking at things, and because that way of seeing is described in terms of determining. So, there is reason enough to discuss the things that are said about a technological view in this chapter, but there is also reason to discuss those things first and independently of the other versions of the determinism thesis.

Heidegger and Fascism

The interpretation of the determinism thesis we explore originates from thinkers such as Ellul and Heidegger. We focus on Heidegger because he is one of the most influential philosophers of the last century, whose impact extended beyond his lifetime. However, choosing Heidegger might be contentious for some, given his affiliation with the National Socialist German Workers Party, the Nazi party. This raises a critical question: should we still engage with his thoughts on technology despite this background?

Indeed, Heidegger's political affiliations place him on the wrong side of history. Yet, this fact alone should not lead us to dismiss his work on technology. Our interest lies in the ideas, assertions, and theories themselves. Are they plausible? Are they accurate? The validity of these ideas should be judged independently of their author's personal history. As we noted in Chap. 1, a theory's correctness is not diminished or enhanced by its association with figures such as Simone de Beauvoir or Barack Obama. The personal morals or political actions of a theorist are irrelevant in assessing the validity of their ideas. Even statements made by a fascist can be true, and they are not false merely *because* of who said them.

Therefore, although Heidegger's political missteps are significant, they do not preclude a critical engagement with his views on technology. The question remains: is the prevalent assumption that we are locked into a technological, instrumental view of the world accurate? There are certainly issues with Heidegger's version of technological determinism, and clarifying what exactly his perspective entails is essential to evaluate its merits.

The Essence of Things

Heidegger's philosophy of technology is part of a stinging critique of the history of philosophy. He speaks only of Western philosophy; non-Western philosophy escapes his attack. Heidegger does not claim that this history starts with Plato in the fourth century BC, but according to Heidegger, philosophy has gone wrong since Plato. Since then, philosophers have been thinking about everything that exists, and especially about human beings. They do so in a specific way: first, by focusing on what something is and is not, and second, by looking for the foundational basis of everything that exists. Let us explain this.

From Plato through Thomas Aquinas to Descartes, philosophers have always wondered about the essence of things. They have long asked themselves ontological questions, meaning that they figure out what something is: 'what-is' questions. Applied to humans, this means looking for a property that is characteristic of all the organisms we call 'people.' In addition, it also means that for more than 2000 years philosophers have been looking for unique characteristics, that is, characteristics that occur exclusively in humans. With Aristotle, for example, this led to his well-known description of man as a rational animal. According to him, all humans have the ability to reason and it is also that ability that distinguishes them from animals that are not human.

The history of philosophy, however, is more than just the search for an essence. Certainly until the beginning of modern philosophy around the eighteenth century, it was not enough for most philosophers to determine what something is. They also thought about something that transcends everything else, that is, something that justifies everything else. Throughout history, that has usually been God; He is the one who justifies and supports everything that exists, from animals to plants. People have their own desires, but ultimately all life exists for the sake of God.

Heidegger wanted to radically break with this type of reflection. Philosophers may have reflected on God or on unique and universal attributes, but that does not clarify the ways in which people live. The question Heidegger therefore put forward, and around which much of his oeuvre also revolves, is not what something is, but how something is. In his view, the core task of philosophy is not to look for an essence, but for existence. How do people stand in life? What is being human?

The Technological Imperative

An important part of his answer to the question of how we stand in life today is that a good part of our everyday doings is not guided by explicit reflection but by a spontaneous understanding of reality. When one has to cut an onion while cooking, one does not first form an image of a knife, of what it looks like and what it is for. No: one reaches for the knife in an unthinking way, because one involuntarily understands a knife as something that serves to cut. This is an example of what is also the case more generally, Heidegger believes. The way in which we live today is driven by a spontaneous understanding, not only of my cooking skills but of reality tout court (Feenberg, 1999).

That concept has a historical character. Heidegger strongly emphasized this. Everything changes: societies, people, but also concepts. The framework from which we view the world today differs from a medieval one. Of course, it is impossible to pinpoint a moment when the pre-modern understanding of reality turns into the modern one, but that does not alter the fact that there is a difference between how we approach the world today and how it was done in the past. In the sixteenth century, a Spanish court could still rule that people should not change the course of a river, and animals could still be held accountable in a court of law. After all, everything is a product of God, and so it would be disrespectful to tamper with God's work. Today, such a judgment is virtually unthinkable, and that is because we no longer understand reality as an effect of an intention, let alone a divine will.

If in the pre-modern era everything was understood in a religious sense, what is peculiar to a modern view? Heidegger, who was influenced on this point by writer Ernst Jünger, believes that today we look at the world through technological glasses. What does he mean by this?

In Chap. 1 we pointed out that we often use the term 'value' in an instrumental sense. Something has such value when it contributes to a goal. Anything that falls under the heading of 'technology' is designed with a purpose in mind and thus, by definition, has instrumental value (provided it is not defective). When Heidegger claims that we conceive of reality in technological terms, he means to say that our era is characterized by an instrumental view of reality. Sport, art, or data, we view these things as instruments, according to him. Sport is a tool to become fit or attractive, art a tool to jolt the world's conscience, and the data we leave on the web serves to feed AI systems or to profit tech companies. We consider things in the world as tools to accomplish our projects, to satisfy our desires, to meet milestones. Thinking in terms of utility is what characterizes our modern age, at least according to Heidegger. Of course, communism and capitalism, two systems of the modern age, are different, but what connects them is an instrumental view of reality.

It goes without saying that designers and AI developers think primarily in terms of utility. After all, their job is to design technologies, things that are a means to an end. Less obvious is that this utility thinking has also penetrated the fields where not everything revolves primarily around technology. According to Heidegger, this is what characterizes our era. Thinking in terms of means and ends not only dominates the tech world, but has crept into all areas of society. We are colonized by an instrumental gaze—to paraphrase the terminology as used by philosopher Jürgen Habermas. From the political world to the cultural sector, from health care to education, the instrumental way of thinking is everywhere. There is nothing that escapes that gaze, man included. Nature is instrumentalized as well. Coal serves industry, wind is used to generate electricity, mountains are used to relax after a busy work year, beautiful landscapes make a selfie on Instagram more appealing, and the Rhine, Heidegger says in a famous passage, is a means to drive the turbines of a hydroelectric power station.

Heidegger goes even further (Feenberg, 1999). One cannot infer from the foregoing that all spheres are governed by a technological way of thinking. For that thinking may have penetrated all spheres, but that does not mean that a technological framework is the only one from which things are viewed. Nevertheless, to revisit the quote from the introduction, Heidegger claims that instrumental thinking hangs over our heads like destiny. Thinking in terms of means and ends dominates all domains and has supplanted all other views. The result is that every human being is in the grip of a perspective from which there is no escape and in which it is impossible for anyone to look at the world in a non-instrumental way. For example, we can now understand nature only as a means, and not as something that has value in itself. Heidegger's philosophy of technology is about the proliferation of the imperative of utility thinking (Couldry & Mejias, 2019).

Technologization and Technocratization

It is clear that at the heart of Heidegger's text are not technologies—airplanes, cars, kitchen robots—but rather a technological way of thinking, a framework, a gaze. Heidegger does not draw attention to our technological condition, the fact that our existence is permeated by technologies. He points to the framework of thought that has been pushed across the world in modernity and to the technological imperative that makes itself felt everywhere and at all times.

Yet Heidegger's philosophy of technology, like Ellul's thinking, is a textbook example of the determinism thesis, or at least of a particular interpretation of it. To understand this interpretation properly, we bring to mind that determinism is about a causal bond in which the effect is necessary, whereby the cause guarantees the effect. Heidegger's technological determinism can therefore be formulated as follows. Our culture is governed by a technological way of thinking—that is the cause. There is a way of thinking that is not owned by any one person, party, or corporation, but that has spread across all domains of society and spans the entire society in such a way that there is no place left for any view other than the instrumental way of seeing. The consequence, then, is that we moderns inevitably think in terms of means and ends, and can only look at the world in those terms. Given that our era is so permeated by an instrumental view, it must be the case that we think from that perspective. This is true when it comes to art or politics, but equally true when it comes to nature or education.

In conclusion, we would like to bring two things to your attention. First, Heidegger does not provide an explanation for the alleged instrumentalism of our time. He describes *what* he believes to be the case, but not *why* this is so or what the cause of the primacy of the technological gaze is. He does believe, however, that this gaze is not the result of our technological condition. That we can think only in instrumental terms is not a consequence of our world being permeated by technologies. Rather the opposite is the case, Heidegger argues. We understand the world only in terms of utility, which is why more and more technologies are made; our existence is strongly tied to technology precisely because we have come to think exclusively in instrumental terms.

Furthermore, you cannot attack Heidegger's thinking by pointing out that moral values are still taken into account in decision-making processes today, in politics, for example, or that groups still have a voice today in order to defend their interests. This is because 'technologization' does not mean the same as what we call 'technocratization.' Heidegger believes that our culture is permeated by a technological way of thinking, and here that means by an instrumental way of thinking. However, that description does not coincide with the characterization of our culture as a technocratic culture. The latter means that decisions are left to experts, people with very specialized knowledge, who are guided solely by facts and not by values or interests.

From Heidegger to Google

Quite a few people are attracted to Heidegger's view. In a way, this is not surprising. After all, there are an awful lot of examples of technological, instrumental thinking. When artists apply for grants, for example, they have to explain what their projects will lead to. Will it serve to raise even more funds? Will it put our country on the map? Will sectors other than the arts benefit? If you cannot answer those questions in the affirmative, you have little or no chance of getting funding. There are also numerous examples of utilitarian thinking in other fields: top athletes pursue selection for the national team because it increases their chances of a transfer; students do internships abroad in order to acquire a better position in the labor market;

researchers give lectures at conferences with the intention of strengthening their curriculum vitae; the teaching of Latin, Greek, and art history in secondary education is criticized, not because it is not interesting, but because it serves little or no purpose, according to some people.

The framework outlined by Heidegger is now several decades old, yet contemporary phenomena still fit into it. Furthermore we, as people—and not just a sample or focus group, but all of us—are seen as a tool. More specifically, our data constitute a tool. The traces we leave on the web are a means of making money. At least this is evident from the business model introduced around 2000 by the founders of Google: Sergey Brin and Lawrence Page, better known as Larry Page.

In 1995 the two met at Stanford University, pretty much the generator of tech people for Silicon Valley. A year later they designed the now famous algorithm PageRank. This algorithm brought order to what had until then been a confusing Internet, by making the search for information more focused. If Page and Brin had left it at that, the world would have looked very different today. Originally, that was the intention. In a 1998 paper, they wrote that Google was primarily a search engine for academics and that advertising did not fit well into that context (Brin & Page, 1998). But they quickly abandoned this idea. This is because they wanted to make Google the search engine, Google the money machine. In 1999, they tried to get rid of the company by trying to sell it to Altavista and Yahoo. But when that failed, Page and Brin decided to switch to targeted advertising anyway: companies buy space on Google's web pages that they can fill with personalized messages. This turned out to be a hit. In 2001, revenues rose from US$19 to 86 million; in 2004, they were making more than US$3 billion in profit. Today, you could think of Google as an advertising company, along with Facebook. An estimated 70% of all online ads are placed on the platforms of both companies. Although, of course, that is not to say that online advertising leads to more consumption—at least this is what the research suggests, as we touched upon in Chap. 2.

In such a business model, the importance of data increases, and we become nothing but a collection of data. Because the more data one has about you—the better one knows what your preferences and desires are, what you want to do and will do—the more sophisticated the algorithms, and thus the better companies or other advertisers can tailor their advertising messages to the potential consumer. That is why data traders such as Experian and Axiom exist, companies that follow us offline everywhere (geotracking) and relentlessly survey our online behavior, every purchase or like, and then resell these data to advertisers. It is also the reason why Google Maps, among others, was developed and why platforms such as Facebook are constantly after our attention and want us online as much as possible. By the way, we now know that they have succeeded on the latter point (admittedly, also from first-hand experience). Anyone with a Facebook account is said to spend an average of 50 min a day on the platform. Instagram and Snapchat have to make do with less: each day the apps are used for 20 and 30 min respectively (Vaidhyanathan, 2018).

So our data are also included in an instrumental logic. Platforms want our data to sell to advertisers; advertisers look for consumers and therefore feed their algorithms with our data. With that, we have yet another example of utility thinking, one that Heidegger himself, more than five decades back, could not have imagined. Does that prove him right? If there are so many cases that testify to an instrumental logic, do they not prove that we are trapped in this logic?

Beyond Utility

Heidegger's version of the determinism thesis suffers from at least two problems. The first has to do with the persuasiveness of the arguments. The second problem is that, despite the many examples of technological thinking, there is also more than one case that contradicts Heidegger's analysis.

What exactly is the first problem? Those who defend determinism set the bar high. The claim is that, given a certain initial situation, a certain effect will necessarily follow. It is then not enough to show that there is a causal connection, it must also be made plausible that it cannot be otherwise than the effect follows from a certain cause. This is true of Heidegger and of the variants of technological determinism discussed later. The problem, however, is that in Heidegger's *Die Frage nach der Technik*, no sufficient reasons are given to conclude that technological determinism is correct. Of course, he gives examples of how we approach things in an instrumental way. He talks, among other things, about the water of the Rhine being used to power a hydroelectric station, and more generally, about nature being seen as a source of energy. But those few examples are not enough to assume that we inevitably think in terms of utility. So our point is not—at least not here—that what Heidegger claims is wrong, or that his version of technological determinism is wrong. The criticism is that his argument for his thesis is flawed. In order to defend the thesis he is defending, he should have provided a firmer foundation.

One possible response to that criticism might be to look for scientific evidence. Although Heidegger—who nevertheless makes statements about how we look at the world today—does not himself refer to scientific studies, one could check whether research shows that instrumental thinking has penetrated all levels of society. Such research may be practically unfeasible, but suppose it were done anyway and yielded these results, would that support Heidegger's reflections?

If something necessarily exists, then by definition it is so in all cases. A square necessarily has four corners, and so it is true in all cases that a square has four corners. If you can point to one instance where it does not exist, then it is not necessary. The reverse, however, is not true: if something is the case in all contexts, then you cannot always infer that it is necessarily so. Suppose you see that it is raining on a Saturday and the following week you also observe that it is raining on Saturday. In a freak coincidence, you see that it rains on all Saturdays of that year. From this, however, you cannot conclude that it necessarily rains on Saturdays. Thus, if sociological or anthropological research were to show that instrumental thinking has spread across all domains of society, then although you might have reason to suspect that it is inevitable, you cannot infer it with certainty. Such scientific findings, in other words, would not eliminate the fact that Heidegger's thesis lacks sufficient support. It would still have to be made plausible that technological thinking is necessary, and not merely something that is present in all domains. Heidegger could have anticipated such criticism by being more moderate and making less strong claims.

In addition, there is a second problem. Not only does Heidegger's argumentation fall short, but his central thesis is incorrect. We are not condemned to instrumentalism; another view is possible. Heidegger's analysis is thus insufficiently nuanced, speaks in terms that are too coarse, and overlooks the differences that are nevertheless obvious.

The fact that the world is still more than a means to our ends today is evidenced, among other things, by the fact that people play, have hobbies, conduct fundamental research, or take care of each other. When you care for someone, it means that you do not regard that person as a means but as an end in itself, or at least not exclusively as a means. It may be that caring for someone makes you feel good, but that is not normally the reason for caring. Caring is usually founded on a non-instrumental view that sees the other person as having value in themselves. Those who now claim that there are people who help others because of the good feeling that helping them gives them may well be right, but this does not undermine my criticism. There is more than one person—and perhaps it is true of the vast majority—who cares for the sake of the person, and that suffices as an argument against Heidegger's determinism. Or take cherishing objects, such as the grandfather's watch from Chap. 1. That object still functions properly, yet that functioning does not absorb its value. The watch has meaning, not only because it shows the correct time, but also, and especially, because that was the last gift the grandfather gave Tom for his birthday. That value will always stick to it, even if it turns out that the watch has become utterly useless.

Some, though agreeing with this critique of Heidegger, believe that the increasing use of technology nevertheless threatens a non-instrumental view. In a sense this is understandable, for technology is by definition made with a purpose in mind. Yet technology need not be a threat to an approach that does not judge in terms of means and ends; the two can coexist. Indeed, some technologies are precisely the expression of a non-instrumental view. We are thinking first and foremost of medical technology, things that are made for the purpose of care and that are based on the idea that people have value in themselves. Yes, medical devices are a means of caring for others. But they fit into the framework of a non-instrumental approach to others. Obvious examples are defibrillators, MRI scanners, pacemakers, infusion pumps and stethoscopes, but also the AI system we mentioned at the beginning of the book, designed to fight infectious diseases.

These examples do not take away from the fact that we often think in terms of means and ends, that it is not unlikely that we do so more than in the past, and that some may have the impression that instrumental thinking is proliferating. And of course it cannot be ruled out, at least in principle, that there will someday be a time when we can only think in technological terms, though we think the probability is zero or very small. Still, those examples show that it is false that the world can only be understood in a technological sense. Now, does that mean that the reverse is true? Are we to infer that it is impossible to understand the world in purely instrumental terms? No. This is evident, among other things, in how we treat many technologies. A laptop is made with all kinds of purposes in mind, and we understand and use it only in that way. Or to take a current theme, our data are often seen by tech giants as a quick way to make money, and nothing else. The fact that the data belong to us and that we should have control over them is often forgotten. By the way: the latter is of course related to economic interests. Looking at data in a non-instrumental way, for example by linking personal data to the right to privacy, is for Big Tech a brake on the pursuit of profit (Foroohar, 2021).

The Problem with Instrumentalization

All forms of technological determinism are descriptive: they represent how the world works—at least that is the purpose. To conclude the discussion of Heidegger, we would like to point out that he also makes a judgment about this. That the instrumental imperative would proliferate is undesirable; it would be better if such thinking were not so prevalent.

Thus, the tone of Heidegger's essay is predominantly pessimistic. At the time of his writing, this was not so remarkable. The works of other early philosophers of technology—Karl Jaspers, for instance—were also mostly negative, and differ in that respect from those of many philosophers of technology who came after them. For the *petite histoire*: that Heidegger's analysis is quite gloomy is also not surprising given his personal preferences. For example, he is said to have been disparaging to his neighbors because they watched a lot of television. And he was reportedly disturbed by a broadcast about the painting of Paul Klee, because the fast camera movements prevented an in-depth reflection on Klee's art (Petzet, 1993).

In an interview with *Der Spiegel* in 1966, he said that only God can save us. Applied to the subject matter at hand, one could interpret that as follows. God can counteract the dominance of an instrumental gaze in our culture. He, and by extension religion or a more religiously inspired view, can break through the domination of technological thinking, with the result that people regain a sense of the value that people and things have in themselves, regardless of their usefulness. This is at least the critical potential that Heidegger ascribes to God and religion. We will not go into whether this solution makes sense. It does, in any case, presuppose that this alleged remedy is correct. What Heidegger puts forward as a solution assumes that it is indeed the case, first, that we are stuck with an instrumental gaze, and second, that the dominance of such a gaze is undesirable. A moment ago, we questioned the first point. The question now is whether the second point is true. Again, we think Heidegger has support. Not only do quite a few people agree with Heidegger that we can only think instrumentally but also a significant number of people, we suspect, agree that it is problematic that we can only see things as means. Is one right in this? Is such a negative view convincing?

For numerous things, it is not a problem to see them as means. All technologies are designed to realize an end, and so it would be absurd to claim that it is undesirable to understand them in an instrumental sense. The same is true of artifacts that are not technologies. Knowledge can be an end in itself, but seeing scientific knowledge as a means of making technology is obviously no problem either, just as it is perfectly unproblematic to listen to music as a means of relaxation. Can we say the same about non-artificial things? According to Immanuel Kant, we should regard every human being as an end in itself and we should not regard anyone as a mere means. But it is important not to lose sight of another Kantian insight: treating human beings as ends in themselves—as rational authors of their own lives—is perfectly compatible with having them fulfill instrumental purposes as well. Consider the mundane example of paying someone to do some work (dig a trench, write a program, whatever): in one sense, we are using that person as a means of achieving some outcome that we want, but provided that we also at the same time treat them as an end—that we ask for, and respect their consent, giving them the opportunity to consider whether working for us helps them to achieve their own goals—we are giving them the due respect they deserve. Thinking instrumentally about someone or something is perfectly compatible with affording them the relevant moral consideration.

So, if you want to criticize instrumental thinking, it is advisable to be careful. This is not a criticism of Heidegger, however, because he does not claim that it is in itself problematic to approach things instrumentally. He does believe, however, that it is wrong to understand them exclusively in this way. And this, of course, is at least partly correct. It is permissible to hire someone to make a profit, but it is unacceptable to think of that person as someone who has no rights and to whom you owe no wages. Although it is not a problem to view cats as companion animals, it is irresponsible to view them as organisms that have no right to food and drink. The same also applies to nonliving things, data for example. Using anonymized data to train medical AI systems is not necessarily objectionable, on the contrary. You can hardly object to feeding AI with data that can no longer be linked to a person in order to better detect cancer. But it is undesirable when medical information is seen only as a means of making a profit by selling it to a data trader. So the problem we are raising here is not so much that a photo on Facebook or Instagram is being used by Big Tech without permission (although that is also problematic), nor that the context of the photo is disregarded. What we are pointing out is that it is a problem that the photo has no meaning or weight, except as a means of training the AI facial recognition system (Buchanan, 2011).

Nevertheless, the question of whether Heidegger's criticism is not somewhat over the top arises. Although it may be undesirable to think purely in instrumental terms, especially in certain cases, is it necessarily wrong in all cases? Is a purely instrumental view undesirable by definition?

We referred to this in the Introduction to the book. To make batteries, a lot of lithium is needed, which is found in mines in Nevada and elsewhere. The problem, however, is that this mining has unintended consequences—it creates, among other things, large pits filled with toxic black mud. In addition, of course, we also know that burning coal to power machinery emits greenhouse gases. So, although these effects of using lithium and coal are clearly undesirable, our question is why reducing both things to tools would be a problem. Certainly, it is impermissible to understand humans and other animals exclusively in an instrumental sense, but what is the problem when it comes to these kinds of materials? You can ask the same question when it comes to the things that are made and powered on the basis of lithium and coal respectively: technologies. Why should we see them as more than a means?

That, at least to us, is unclear. In addition, it is unclear what Heidegger's answer to this would be. In his text, he gives no arguments for his claim why a purely instrumental view is wrong in all cases. For example, we do not know why he disapproves of seeing things such as drills and cars only as means. Yet this does not mean that in general, apart from Heidegger, no arguments for such a view are given. For example, if we focus on technology and AI, three reasons are usually put forward by others in this context. Are these sufficient to conclude that an exclusively instrumental view of technology and AI is always problematic?

First, some point out that technologies always belong to someone, or at least are owned by someone. As a result, you cannot see them merely as a tool; you also have to treat them with respect and care. This objection may be valid when it comes to the property of others, but what if it is about your property? The second answer relates only to AI. Reducing such technology to tools is rejected because smart technology would have rights. This is not the place to examine that proposal in depth, but suppose it makes sense. Then that too is not enough. After all, not all technology is smart technology. Hence again our question: why would it be wrong to see dumb things like a bicycle or coffee maker as means to an end, and merely as means to an end? Finally, a third answer is that it is not good to view technologies exclusively as tools, because by doing so we reinforce purely instrumental attitudes towards people, or at least there is a high risk that this will happen. Or for those who think that Heidegger is wrong and that today we do not see people only as means: it is undesirable to view technology exclusively as means, because by doing so we no longer attribute value to other people in themselves, or at least there is a great risk that we will not do so. Thus, it is not claimed here that a purely instrumental view of technology is problematic per se. Such a view, according to this approach, is only a problem because it has undesirable consequences; in this case, the instrumentalization of people, the enhancement of interpersonal instrumentalization, or the significant risk of either of these things happening. Clearly, this in itself is not enough to problematize a purely instrumental view of technology. To be convincing, further support is needed, preferably coming from empirical research. Is it indeed true that an exclusively instrumental view of things is being transferred to how we interact with people, or is there a great danger of this? Does a particular view of technology have an undesirable effect on our view of people? Applied to a current debate, for example, is it undesirable to view a sex robot as merely a means of satisfying sexual desires, because the consequence is to reduce other humans to pure objects of desire, or because there is a high probability that this will happen?

With that, we can now conclude the section on Heidegger, on the first version of the determinism thesis. Our conclusion is as follows. The claim that our thinking is determined is not correct. It is not the case that we can only look at the world in a technological sense. But even if Heidegger were right, it is a problem that he does not explain why that would be undesirable in all cases, i.e., not only when it comes to humans and nonhuman animals but also when it comes to, say, coal or technology. To be clear, we are not claiming that there are not good reasons for a general rejection of a purely technological view, just that we do not see them. In any case, the three arguments given by others in the context of technology and AI are not sufficient. For example, it is not at all certain that concerns that an exclusively instrumental view of technology might lead to such views of people are justified.

Types of Technological Determinism

Earlier, we highlighted that beyond Heidegger's perspective, there are three additional variants of the determinism thesis commonly discussed among technologists. Although all these theories share Heidegger's basic premise of a causal relationship—because determinism inherently involves cause and effect—they diverge significantly in their specifics. Unlike Heidegger's focus on a philosophical, technological view, these variants deal more concretely with actual technologies such as robots, laptops, and telephones—tangible, purpose-built creations.

Among these variants, one specifically addresses the seemingly inevitable social consequences of technology, which we set aside for now. Instead, we focus on explaining the other two: one pertains to the emergence of technology, and the other to the evolution of technologies after their initial introduction. Our aim is to delineate these latter two interpretations clearly.

Emergence and Development

If you focus on the interpretation of the determinism thesis about the origin of technology, then the thesis is that all technology exists necessarily. Every technology, in that case, is a necessary consequence of some process or other that preceded it. It could not be otherwise than for technology to be developed. It may seem that computers and social media need not have been invented; on closer inspection they are the inevitable result of, say, World War II and the process of individualization that has been going on for several decades. The wheel and the Internet, they had to be created at some point.

If you concentrate on the evolution of technologies, then it may be about the succession of different technologies or about the renewal of existing technology. Here, it is always about the inevitable link between two technologies. One thing necessarily leads to the next; under all circumstances this technology would lead to the development of that other technology. An example of this form of technological determinism is the following claim. It was inevitable that at the end of the eighteenth century, engineer Claude François Jouffroy d'Abbans would build the first steamboat—the so-called Pyroscaphe—because boats and steam engines already existed—boats were already used in antiquity and the first steam engine was invented by Thomas Newcomen at the beginning of the eighteenth century. That line of thinking also characterizes the 1992 documentary about the history of the computer, *The Machine That Changed the World*. The guiding idea of this documentary is that each step in the development process had to lead to a breakthrough.

There are obvious differences between the two versions of the determinism thesis, between the one about the origin and the one about the evolution of technology. One version deals with one technology, the other with at least two technologies. One version asserts that all technology is necessary, whereas the other asserts that all technology inevitably leads to other technologies. The latter claim implies that some technology is necessary—namely, the technology that inevitably follows from another—but not that all technology is inevitable, which is what the other version claims.

Yet we also want to emphasize the similarity between the two forms. Both when talking about the emergence of technology and when talking about its evolution, it is claimed that technology will come to exist in all cases. This means that it is not a result of deliberation or compromise among all kinds of groups, that it is not an effect of social processes, relationships, or groups. If it were, then technology could not exist either, because relationships between people and groups are not set in stone, just as compromise is not. In addition, the inevitability of technology also means that governments and other stakeholders are passive about its creation and further development. You can interpret this in the following cultural historical way. From the Enlightenment onward, various things were subject at least in part to social, democratic control: ethics, economics, politics. Technology, however, is perhaps one of the last things to escape this. God is dead, most adherents of technological determinism agree. But that does not mean that all transcendence has disappeared. Technology has taken the place of God, one could say of the determinism thesis—at least a certain interpretation of that thesis. Whereas it was once God who eluded the grasp of the people, now it is technology that transcends human control.

Don't Shoot the Messenger

Later, we highlight the arguments for and against both forms of technological determinism. However, we want to stress right now that such reflection has practical relevance. If it is true that the emergence and evolution of technology are inevitable, then this has significant consequences. Governments and other stakeholders would then have no role to play in the design and production of things. If they can exert any influence at all, then it is reserved for the post-production phase. This does not mean, of course, that their influence is negligible. Politicians and groups can be instrumental in spreading technology among the population and in controlling citizens' access to technology—often meaning broadening access, or narrowing it, as in the case of military technologies, for example. In addition, engaged citizens may seek to influence the beneficial or harmful effects of technology. If it turns out that a technology has undesirable effects, governments can anticipate them by trying to influence them; desirable effects can then be reinforced.

There are, in addition, other good practical reasons for taking a closer look at both forms of determinism. These have to do with the kind of interest people may have in technological determinism. In the previous chapters we pointed out that you can have an interest in assertions from multiple angles. You can have an interest in a purely theoretical sense, but there may be other interests involved as well. This is also true of the determinism thesis about the origin and evolution of technology.

On the one hand, the interest may be purely theoretical, meaning that you develop arguments exclusively for or against technological determinism. On the other hand, it may be that some, primarily technology, developers, defend determinism about the origin and evolution of technology because they want to defend themselves against criticism of the technology that they have developed. You may be opposed to that development, they defend, but that criticism is parried by the fact that the technology will come anyway. The defense of the determinism thesis is used here to marginalize negative comments. This, of course, makes it relevant to delve further into technological determinism. If it turns out to be correct, it provides the tech world with ammunition; if, on the other hand, it turns out that determinism does not hold water, it deprives technophiles and developers of a possible response to criticism of their designs.

Another possible motive—again, mainly from tech designers—is of a moral nature: one wants to wash one's hands clean of the problem. People are interested in this theory because they are convinced that determinism means that you cannot be held responsible for the technology. The reasoning then goes like this: if one designs a technology, a technology that was going to exist anyway, then one cannot be held morally responsible for the negative consequences that may arise from the use of that technology. We return to this point; but suppose for the moment that determinism does indeed absorb one's moral responsibility. The consequence of this is that technology developers have no moral responsibility and so cannot be punished for undesirable effects. For this reason alone, it is relevant to take a proper look at determinism later on.

To avoid confusion, we want to emphasize that these human motives for being interested in determinism are not a reason to attack or even reject the theory, but rather to examine it more closely. It is possible that determinism is wrong, and it is possible that the theory stems from self-interest. But if the determinism thesis is wrong, it is because there is something wrong with the theory, and not because the defense of the theory stems from certain motives. We need to focus on the theory, not on the person and the possible interests behind the theory.

In the Grip of Experts

In addition to Heidegger's version of technological determinism and the two versions of the history of the development of technology, there is a fourth, final, version, which we introduce here. It states that all technology inevitably has social effects; those effects would be there in any context. Even if we went back in time and introduced the technology elsewhere, the exact same social effects would occur.

In a way, this claim is similar to the disruption thesis from Chap. 2, the claim that AI is disruptive technology. Both focus not on technology per se, but on the consequences of using technology. Another similarity is that the two cases are about technology, rather than a way of thinking. Moreover, they now zoom in on a causal relationship in which technology itself is not an effect, but a cause of something else. Yet there are also clear differences between the disruption thesis and the social effects thesis. These differences have to do with the effect of technology. The disruption thesis, as we have interpreted it, refers to the impact of AI in terms of ethics; the original interpretation is economic in nature and deals with the effect of AI on the market. But although the disruption thesis can in principle be about consequences in all areas—about economic and moral but also social consequences—in the determinism context the focus is exclusively on social effects, or the consequences of technology on social relations, processes, and roles. Furthermore, whether the social effect is disruptive or not is not important for the determinism thesis.

Usually, this interpretation is not mentioned, or it is discussed separately. One tends to link it to two other meanings of the determinism thesis, those about the emergence and evolution of technology. It cannot be otherwise than that technology comes into being and that from one technology comes another, and so it cannot be otherwise than that technology exerts social influence: so the thesis goes. If this is true, then it means, first, that groups of people, social relations, roles, and processes in society are influenced by the decisions of a handful of experts in the field of technology—designers, computer scientists and AI developers who strive for ever more efficient and profitable technology and who rely on the exact sciences and engineering to do so. Second, it also means that the lives of many people are affected by technologies over which they have no control. If the combination of the three forms of determinism is correct, then we need no further clarification to see that this is an undesirable situation. But is this indeed true?

In a sense, it is understandable that determinism concerning social effects on the one hand and determinism concerning origins and evolution on the other are linked. The idea of social impact seems to follow seamlessly from either of the other two forms of determinism. Should it turn out that the development of one technology necessarily leads to other technologies, it is at first glance difficult to imagine that development without social effects; or, at least, it would not be surprising that the unstoppable progress of technology also has social impacts. Yet these different interpretations are not inextricably linked. We have little trouble imagining that technologies inevitably succeed each other without those innovations having social impact. And conversely, it is not too difficult to imagine a world in which the emergence or evolution of technology has social effects, whereas that emergence or evolution does not have an inevitable, necessary character.

So, these interpretations are separate, at least in theory, and consequently must be discussed separately. We emphasize this because those who link these interpretations usually argue only for the interpretation about the emergence and evolution of technology. One seems to assume that the reasons for interpreting the determinism thesis about developmental history also apply to the interpretation of social effects, but this is thus unjustified. Hence our question: is it true that technologies have effects that are both social and inevitable? This is the question on which we are going to focus now. In the next few pages we evaluate the version of determinism that concerns social effects. It is only toward the end of this chapter that we highlight the arguments for and against the versions of the determinism thesis that have to do with the emergence and evolution of technology.

The Social Effects of Technology

Because, by and large, little attention is paid to determinism with respect to social effects, it is important to highlight some common reasoning errors. We do so by referring to the following two issues: on the one hand, social media and the algorithms that control the forums of those media, and on the other, the polarization of society. The latter characterizes several countries. We know that in the USA, for example, society is increasingly divided into Republicans and Democrats. The first group argues against immigration and is in favor of gun ownership; the second group is pro-immigration and opposed to gun ownership. As an aside, Levi's jeans are reportedly worn more by Democrats than by Republicans.

Are Social Media Polarizing?

It is often argued that the two are causally linked: social media encourage polarization. The reasoning then goes as follows: social media have existed since the first decade of this century and since then polarization has increased, so the conclusion must be that polarization is a consequence of social media. Let us assume for a moment that the chronology is correct: first social media, then polarization. However, that does not justify the conclusion that social media contribute to polarization. It is not because two events follow each other temporally that the event that came first is the cause of the second. When cyclist Alejandro Valverde prays before riding in a race and he wins the race afterward, it does not mean that he won because he prayed.

Others, in turn, argue this way. The use of social media is associated with the increase in polarization, and so technology is the cause of that polarization. That conclusion, too, is unjustified, even if it is true that there is a correlation. This is where the confusion between a correlational relationship and a causal relationship comes into play. The fact that two phenomena are correlated does not imply that there is a cause-and-effect relationship. For example, it is true that the increase in crime rates correlates with the increase in ice-cream sales, but that does not mean that there are more crimes because more ice cream is sold, or vice versa. The increase in both is caused by a third factor: the increase in temperature in the summer.

To be clear, we are not claiming here that social media do not effectively promote polarization. It is possible that they do. We are just claiming that you cannot conclude this based on the following data: that polarization over time follows the use of social media and that there is a correlational relationship between the two phenomena.

Not only must the determinist beware of sloppy reasoning, but those who attack it must also be vigilant against hasty conclusions. This is evident, for example, in this case. When thinking about the influence of social media on polarization, scholarly studies such as the 2019 paper 'The Editor vs. The Algorithm' and the work of economist Ro'ee Levy are often invoked (Claussen et al., 2019). These studies support the idea of social media as an echo chamber. Algorithms lead to a decrease in the diversity of articles shown in the newsfeed on Facebook, for example, and over time people spend less and less time reading different types of articles. Furthermore, it is also true that algorithms often present us with articles that primarily confirm our beliefs, rather than challenge them. In short, there is evidence to support the claim that social media create ideological bubbles.

On the other hand, scholars also point out that there is no evidence proving the claim that algorithms and social media contribute to the polarization of society (Levy, 2021). They may create echo chambers, but that AI systems increase tension and antagonism between groups has not been demonstrated to date. Some could draw the conclusion on this basis that there is no causal relationship between AI and polarization. This too would not justified. If you do not have proof that something exists, that does not prove that it does not exist (Agar, 2015). So even if we have not been able to establish a causal link yet, it is still possible that the link is there, but that we simply have not figured it out yet.

Furthermore, one cannot attack the version of technological determinism about social effects by invoking a study that adequately explains polarization without having to refer to social media as the cause. After all, anyone who is a determinist when it comes to social media and polarization only claims that social media necessarily lead to polarization, not the reverse, namely that all polarization necessarily results from social media. By extension, should scientific research show that polarization arises from social media, but not solely from social media, that is not an argument against technological determinism. Those who think it does are starting from a false understanding of that determinism. The determinist suggests that technology inevitably has social effects, not that those effects arise solely from technology.

Call Me, Write Me

The previous paragraphs are merely a plea for caution when considering the social effects of technology. They do not show that technology has no social effects. It would also be at least surprising should we defend the latter. After all, there are countless examples of how technology affects social processes, roles, and relationships. That influence can be desirable, undesirable, or neutral. The microwave has as its effect that family members eat together less often; cars underline social status; and social media make it easier for us to renew old friendships or make new contacts.

Another example is the use of AI by the legal system and the police. In Chap. 1, we introduced the term 'algoracism' to draw attention to discrimination by AI based on skin color and ethnicity. In that same context, we also referred to the COMPAS algorithm, which has been used in the USA to estimate the likelihood of recidivism. In 2016, it was found that the algorithm treated white people and people of color unequally: the former were released at a quicker rate than the latter. The use of AI by police departments also has its social consequences. When such technology is used to predict where crimes might take place in the USA and the police rushes out to patrol a hot spot, you would be better off being a person with white skin. According to figures from the US Department of Justice, if you are a person of color, you are more than twice as likely to be arrested than if you are white. People with a dark skin color are five times more likely to be stopped without a fair reason than people with a light skin color. These cases illustrate the uncomfortable truth that technology can underscore or even reinforce not only male but also white hegemony (Heaven, 2020).

Another example: the telephone. Because it derives from the telegraph, the telephone was used shortly after the end of the nineteenth century only for short business messages. That changed in the first half of the last century. Partly because of the reduction in call rates, the telephone became increasingly sought after; then, during the interwar period, it was also used as a means of social communication. This shift clearly had a social impact. One initial finding is that not only the number of acquaintances but also the number of conversations with the same person increased. Furthermore, the telephone would also have had an impact on social practices. Some practices, such as short unannounced visits, decreased; rituals such as the announcement of a visit were then reintroduced by telephone. Finally, the number of physical encounters also increased, which may be because the phone allows you to make appointments more quickly. This goes against that popular criticism that new communication technologies lead to less physical and social contact.

Again, the point is not to draw quick conclusions but correct ones. After all, the meaning of the determinism thesis that we are now discussing is about social effects that are inevitable. The cases we have just cited only pointed to the social nature of the effects of technology and say nothing about the inevitability of those effects. Therefore, you may not conclude from the examples given that the social effects could not have been there, just as the conclusion is not justified that the effects are necessarily due to technology.

Necessary? Possibly!

Now suppose that all technologies have social impacts. Would those impacts be there in any society and at any point in time? Do the impacts necessarily follow? Is it enough that technology is there for social impacts to follow? Are utopians and alarmists right when they say that a technology inevitably has good and bad social consequences respectively?

There is a danger in answering this in the affirmative because you made the correct prediction that the social effects would follow. One predicted that the effects would be there, and because the effects have occurred, the impact is inevitable, one might reason. Although it is understandable to draw that conclusion, it is not justified. When something is determined—read: when an effect necessarily follows—only then can you predict in advance that the effect will follow. The reverse is not true. Tom's wife predicts that he will be anxious tomorrow when he is sitting in the dentist's waiting room. When that is indeed the case the following day, it does not mean that it is necessarily so. She may have had good reasons for this—after all, Tom is usually a bit of an anxious person when it comes to the dentist—but it is also possible that it was different this time, for example, because, without her knowledge, Tom attended a number of sessions with a behavioral therapist to control his anxiety.

Furthermore, it is also the case that the effects of technology in areas other than the social domain are not necessarily inevitable. Take a car with a gasoline engine. If you use such technology, it will emit greenhouse gases but will not necessarily create smog. Smog is only created when there are many petrol cars in a small area. Thus, smog is a consequence of driving the car, but not an inevitable consequence. Not only ecological but also social effects do not necessarily follow in all circumstances. That is what the telephone teaches us again. It could only lead to the broadening of interpersonal contacts because it became affordable over time and because people were made aware of its existence. In addition, its success became possible only because the initially negative image of the telephone faded, because production and distribution increased, because many switching centers were built, and because the number of cable connections increased. In a world other than the one in which we live today—with no cheap subscriptions, no cable connections, few distribution options—the telephone would not have had these social effects.

Thus, if technologies have social effects, they are not per se necessary. It is not always enough that the technology exists for the social effects to follow. A number of conditions must be met before those effects can occur, conditions that can be of many kinds: psychological, social, economic, political, material, and so on. This is true not only for the telephone, but also for numerous other technologies, perhaps even all of them. AI systems, for example, have social effects in addition to economic ones, but those effects can only be there when certain conditions are met. There are no social effects of smart systems if the shipping industry does not distribute computers and cell phones worldwide in shipping containers, if the tariffs for users are too high, if the mobile networks do not work, if there is insufficient technological know-how among users, and so on. So again, you have to say: in a different world or a different era—without, for example, international transport—AI would not have the effects on a social level that it has today.

Let us assume, however, that our reasoning is flawed: if technology has social effects, surely these effects follow under all circumstances, irrespective of place and time. May we then conclude that the determinism thesis is correct—or, at least, this interpretation of it? This is highly doubtful. All technologies, by definition, are designed with a purpose in mind, and thus have an effect when they do what they are made to do. Yet, it is far from certain that all technologies have an impact on a social level, let alone that it would be necessary. Consider the drill from Chap. 1 or the electric toothbrush from Chap. 2. It is extremely difficult, if not almost impossible, to conceive of effects on social relationships, processes, or institutions that would follow from the use of these two technologies. It cannot be excluded that even such basic things have a social impact, but it would not be surprising should it turn out that there is none.

We conclude. Not only is Heidegger's version of technological determinism incorrect but also the determinism thesis concerning social impacts suffers from several problems. Although it is not certain, but very likely, that some technology has no social effects, it is clear that not all social effects of technology are necessary. A less strong but, importantly, a more correct thesis is that all technologies have the potential to have such social consequences. That conclusion leaves room for the statement that it is very likely, but not necessary, that, say, Instagram's algorithms have social consequences and that those consequences are more likely than the consequences of, say, an electric toothbrush. And further, it also allows that some technologies inevitably have social impacts. Should it turn out in the future that some technologies do indeed have social impacts anyway, that fact would not contradict our conclusion. But as we stressed earlier: in order to make this plausible it is not enough to just point to the social effects; it would also have to be made clear that these effects are inevitable.

The Birth of Technology

It is time to look back for a moment. This chapter is about a popular thesis that, like the neutrality thesis and disruption thesis, is often defended by politicians, philosophers, computer scientists, and engineers: the determinism thesis. There are at least four versions of this, and by now we know that Heidegger's version and the one about social effects are incorrect. But there are two other versions, which we have already mentioned but not yet evaluated. One is about the emergence of technology, the other about the evolution of technology once it is put on the market. We pointed out earlier that the version of determinism about inevitable social effects is often mentioned in the same breath as the two versions of technological determinism about the emergence and evolution of technology. However, it is important to keep these apart. If it is wrong that technology has inevitable social effects, it does not necessarily mean that the other two versions are also wrong. So, it would be wise to take a closer look at those other versions as well.

Before discussing the specifics, it is important to recognize that examining the origins and development of technology often parallels the way in which we understand nature. This approach gained momentum in the latter half of the nineteenth century, spurred by the burgeoning success of evolutionary theory. For instance, Samuel Butler, in his 1863 essay 'Darwin Among the Machines,' argued that the development and evolution of both organisms and machines should be viewed through Darwinian principles. According to Butler, just as organisms evolve by adapting to their environment more or less effectively than their competitors, so do machines. This perspective is still advocated in contemporary works such as Kevin Kelly's *What Technology Wants*.

Contrasting sharply with this naturalistic view is the eighteenth-century philosophical and scientific perspective that described nature using technological terms—consider the metaphor of God as a watchmaker and the cosmos as a finely tuned clock. However, those who might find the analogy between technology and nature odd or implausible should consider that many technologies are directly inspired by natural forms. A historical example is barbed wire, invented in late nineteenth-century Illinois, which mimicked the strong thorns of the Osage tree (Basalla, 1988). A more modern instance is the CRISPR/Cas9 gene-editing technology, heralded as the breakthrough of the year by *Science* magazine in 2015. This technology models a natural defense mechanism in bacteria, which use molecules such as Cas9 to protect against viruses (Arthur, 2011).

Twice the Wheel

We begin with the interpretation of the determinism thesis about the origin of technology; then, we focus on the evolution of technology, more precisely on the interpretation concerning the succession and innovation of technologies.

All technology had to be invented—this is what we are talking about now. Given the circumstances, it could not be otherwise: all technology arises from something that was not itself a technology. The fact that a technology comes into being is necessarily so. Negotiations and discussions between groups of people cannot change that. That is the determinist's thesis about the emergence of technology. You often hear it, especially among people who are closely involved in the creation of technology. But is it true?

To challenge the idea that all technologies arise out of necessity, critics often point to examples of technologies that seem to emerge independently of such needs. One notable example is the wheel, specifically its invention in Mexico and Central America after the fourth century BC, long before the Spanish colonization and without knowledge of its earlier invention in Europe and Asia around the fourth millennium BC. The motivations behind the invention of the wheel differed significantly across regions. In Europe and Asia, the wheel was developed to transport large and heavy objects that were too cumbersome for people or animals to carry. However, in pre-colonial America, the wheel was used for miniature objects, such as toys or sacrificial items, not out of any practical necessity (Basalla, 1988).

Similarly, the automobile, which emerged at the end of the nineteenth century, was not invented because of a shortage of horses or an existing need for motorized vehicles. Instead, the demand for automobiles developed after their invention. These examples suggest that technological development might be able to occur independently of immediate needs, challenging the deterministic view that technological innovations are always a response to necessity.

Some critics might argue that those examples are a problem for determinism, at least for the interpretation of it that we focus on here. After all, that theory is about the necessary nature of the emergence of technology, whereas the examples just show that technologies are not always created out of necessity. Although we also believe that this type of determinism is insufficiently grounded—more on that in a moment—this attack misses its target. This is because 'necessity' is understood differently each time. Both the critique of determinism and determinism itself are formulated in terms of necessity, but the same term is interpreted differently in the case of the critique than in the case of determinism. Although 'necessity' for the determinist means 'inevitability,' the critique uses 'necessity' to refer to a need, the fact that one needs something, that one has a need for something. In other words, the critique misses the point. This would not be the case if the determinist said that technology comes from a need; but that is not what the determinist claims.

Simultaneous Evolution

Does that mean that this version of determinism does, in fact, hold up? Is it inevitable that a technology will be brought into the world? That depends on the strength of the argument for this theory. The argument that technology did have to arise is based on the idea of simultaneous evolution—and as far as we know it is the only argument. It means that something arises in different contexts even though there is no connection between them. We suggest taking a closer look at this.

The phenomenon occurs in the organic world. Antifreeze, which keeps the blood flowing even when it is very cold, was developed not once but twice: once by fish at the South Pole, once by fish in the Arctic Ocean. Some other examples: navigation by echolocation is used not only by bats but also by dolphins and the Asian palm swift; floating swim bladders are found in mollusks and jellyfish; and in addition to frogs, chameleons have developed specialized tongues to trap their prey from a distance. The simultaneous emergence of a trait also occurs in the plant world. In North America, mushroom genera developed fungi independently of each other in different places, and seven species of plants became insectivores independently of each other in order to acquire sufficient nitrogen (Kelly, 2011).

Simultaneous evolution is a phenomenon that also occurs outside of nature, for example, in the realm of science. In 1979, a gene coding for the so-called protein p53 was discovered. That was a very important discovery in the fight against cancer, because the gene has a negative influence on cell division and thus prevents the development of tumors. This discovery was made in different places and independently of each other: in London, New York, and Paris, among others. In addition, and more importantly, there is also simultaneous evolution in the tech world. In other words, it is often unfair when books or websites on the history of technology link an invention to just one person. Let us have a look at some examples.

The blowgun was invented twice: once in Asia, once in America. It is true that Thomas Edison invented the incandescent lamp in the USA at the end of the nineteenth century, but it is also true that the lamp was invented in England and Russia during the same period, by Joseph Swan and Alexander Lodygin respectively. On 14 February 1876, both Alexander Graham Bell and Elisha Gray applied for patents for the telephone, which they had created separately. And although we primarily associate Bell with the telephone, Gray even filed his application 3 h earlier. Indeed, in 1860 Antonio Meucci had already obtained a patent for the telephone in Italy, but, partly because he was not proficient in English, his patent was not renewed in 1874, 2 years before Bell and Gray. Often, Louis Daguerre and France are linked exclusively to the birth of photography. This, too, is unjustified, for in England at the same time and independently of Daguerre, William Fox Talbot discovered the same principles that formed the basis of photography. The inkjet printer was invented twice during the same period: both in the laboratories of Canon in Japan and by technology company Hewlett-Packard in the USA. A final example: the transistor was invented shortly after World War II both in the USA in the Bell Labs of the telephone company AT&T, and in Paris by two German physicists (Kelly, 2011).

Coincidence and Necessity

For some, these findings on the history of technology are not just fun facts. People also cite them as an argument for determinism about the origins of technology. They reason: now that it turns out that technologies arose independently of each other during the same period, we can conclude that it cannot have been otherwise. Those technologies were bound to arise. There is necessity in the development of the light bulb, telephone, photography, inkjet printer, transistor, and other technologies. Does the argument justify the conclusion?

The first problem is that not everything arises in multiple places simultaneously and independently of each other. In the animal world, the bombardier beetle is unique: it is the only organism that can combine chemicals into a toxic jet that can be sprayed on hostile animals (bombardier beetles also release winds as high as 100 °F). But even in the world of artifacts, not every design is an example of simultaneous evolution. Certainly, many technologies emerge at the same time and independently of each other, but the same is not true of all technologies. The computer, the airplane, and the passenger car: those technologies were not invented simultaneously in different places. Thus, if it were true that simultaneous evolution demonstrates that technology could not but arise, then you cannot generalize the necessary emergence of technology, at least not on the basis of simultaneous evolution.

However, does simultaneous evolution demonstrate that something is necessary? The answer to that is in the negative. That similar things arise in parallel to each other may be a coincidence; one can imagine a world in which it is a coincidence that things arise simultaneously but independently. If it turns out that something arose in different places with no connection between them, it does not follow that it must have necessarily arisen. Thus, one cannot say that simultaneous evolution is a conclusive argument for determinism. Nevertheless, simultaneous evolution is a reason to suspect that it is not a coincidence that something exists. Now, suppose this following conjecture is correct: the simultaneous emergence of something in different places is not a coincidence. Does it then follow that it is necessary?

Statements can only be true or false. There is nothing in between: either a statement is true, or it is false. This is different in the case of necessity and coincidence. The two are not opposed to each other in the same way that truth and falsity are. You must understand this as follows. When something is necessary, it means that it is inevitable. If something is not necessary, then it might as well not have been there. Now, focus on chance, and it turns out that everything that is by chance is not necessary. But—crucially—when something is not accidental, it does not mean that it is necessary. Take, for example, the fact that soccer and cycling are Peter's hobbies. This is obviously not coincidental, because he grew up in a country where both are the most popular sports. But that is not to say that it is necessary for soccer and cycling to be his hobbies. It is possible that he focuses on tennis, for example, because he is good at tennis and not good at soccer and cycling. In short, if it turns out that different technologies emerge simultaneously and independently of each other, you cannot infer that the emergence is also necessary.

It is important to be clear about our point. We are not claiming in the previous paragraphs that the emergence of technology is not necessary, or that there is social control over technology. We have not shown at all that every technology ever designed is contingent. Our criticism here is only that the argument based on simultaneous evolution is flawed, and that you cannot defend determinism based on that argument. Simultaneous evolution does not prove that anything is necessary, even if it turns out that evolution is not accidental. But even if you could infer necessity from simultaneous evolution, you could not do so for all technologies. After all, not everything was designed at the same time and independently of each other.

The Evolution of Technology

A quick recap. We started with Heidegger and then zoomed in on the social effects of technology. Then we focused on the version of the determinism thesis that deals with the emergence of technology. So now it is time to turn to the other aspect of the history of technology. This is about the evolution of technology, about the succession or renewal of technologies once a technology was developed. This is also the fourth and last example of technological determinism.

The central idea is this: technology inevitably leads to the renewal of an existing technology; it could not be otherwise than the fact that a new kind of technology results from an older technology. An example of this form of determinism can be found on the back cover of *The New Digital Age* from 2013 by Jared Cohen and Eric Schmidt (the former CEOs of Google): "But complaining about the inevitable increase in the size and scope of the technology sector distracts us from the real question. Many of the changes we are discussing are inevitable. They are coming." (cited in Zuboff, 2019). Not surprisingly, these phrases come from the former head of a tech giant. That products will lead to new products anyway, that there is an unstoppable succession from old to new technologies is not only a belief that has lasted for a while, it is also one of the creeds of the tech industry in general and the AI sector in particular.

The QWERTY Keyboard

Later, we focus on the arguments. Is this version of the determinism thesis correct or not? First, however, we want to address the explanation given for technology's inevitable progress. This explanation should be understood as follows. Technologies are like planets—that is the central claim. They move forward inevitably and are not moved by collective interventions and democratic deliberation among groups. We have seen this before: this form of determinism also corresponds to the version of determinism concerning the origin of technology on that level. In this context, however, what drives technological progress is usually delineated. Two things are commonly put forward. The first explanation is the pursuit of the improvement of technology. This improvement is not about ethics, and instead must be understood in a technical sense. It refers to the desire to make technology more efficient or profitable. The process of inevitable technological development, it is claimed, is driven by the desire to minimize energy and time expenditure. Second, there is a reference to scientific knowledge. More efficient or cost-effective technology can be made, but innovation, of course, does not arise ex nihilo. It is based on science. It is through scientific understanding that one is able to make new, more efficient or cost-effective technology, to bring technology to market that is increasingly technically ingenious.

Is this convincing? Assume for a moment that it is true that technology inevitably results in new technology. Does that also mean that the explanation given for this is correct? Is all innovation indeed driven solely by the pursuit of technical improvement by a limited circle of hyper-specialized designers and computer scientists?

Take the QWERTY keyboard on computers, named after the order of the first six letters on the top left of the board. This letter order dates back to the days when computers did not exist and when typewriters were still used. The purpose was to slow down the fast typing of typists. If one typed too fast, the mechanical letter hammers would collide and block. This was prevented by the QWERTY keyboard. By placing these letters at the top left of the board, the typists would have no such problem. The technical, practical consideration to opt for such a board lapsed when the switch was made to typewriters without hammers: computers. Nevertheless, people continued to make QWERTY keyboards even for that new technology—and such boards still exist today. This had everything to do with habit. Because typists were used to working with such a keyboard, the design of computers was adapted to that habit. Conclusion? The past can cast a shadow over technologies, so a new design is not necessarily driven solely by the pursuit of efficiency.

Let there be no doubt about this: we are not claiming that there is no technology development that can be adequately explained by the desire for efficiency. Nor is our point that in the cases where efficiency is not sufficient as an explanation, as in the keyboard example, that value then plays no role in the design process. Our point here is simply that not every innovation is explained solely by the desire for efficiency. To believe that every technological innovation is driven solely by that desire is to take too much of a one-sided approach. It is a view of technology that may be distorted by the perspective from which technology is looked at, which in this case is the framework of expert engineering or computer science. In other words, there may be other nontechnical issues involved in the development process, such as ethics, for example, even when it appears that there is an unstoppable progression from one technology to another. And perhaps those other issues come from people who are not AI developers, computer scientists, or designers: they might come from governments, unions, social movements, and so on. You can think of a collective that stands up for the rights of people with physical disabilities, for example, and that fights for a design to be accessible to them, even if that would mean that the technology is less efficient.

The Steam Engine

What about the other explanation? Is innovation always based on science? Does making technology more efficient or profitable in all cases start from scientific insight derived primarily from fields such as the exact sciences and engineering? Although this is often the case, it is not always so.

The belief that new and more efficient technology is based on science is reminiscent of the popular idea that technology is applied science—just as music would be applied mathematics. This view originated in the mid-nineteenth century and was increasingly shared in the decades that followed. It was at the root of the famous motto of the 1933 Chicago World's Fair—Science Finds, Industry Applies, Man Adapts—and was defended in 1966 by philosopher of science and physicist Mario Bunge in a text with the unsurprising title 'Technology As Applied Science.' The central idea is that all technology is based on scientific knowledge, especially knowledge from the natural sciences. Moreover, all technology necessarily starts from science. There would be no technology at all if there were no science. There must first be knowledge that comes from scientific research and is then used to design technology. No technology without science, just as according to the three monotheistic religions there is no life without God, and just as it is impossible to dream or ruminate without a brain.

At first glance, this seems correct. Numerous technologies could not have existed without science: no telephone without knowledge of electromagnetism, no AI without computer science, and no space shuttle without physics and chemistry. But as we wrote in the Introduction, technology has been around for a long time, and more importantly, technology is older than science. The Inuit in northern Canada and Greenland, for example, lived somewhere around the second millennium before our era and made parkas, boots, gloves, igloos, knives, and kayaks in order to cope with the harsh polar climate. And one of the oldest forms of technology dates back to 30,000 BC: the bow and arrow. These items were made without any kind of scientific knowledge.

Some may find this unconvincing. Are clothes and bow and arrows technologies? We remind you that, following in the footsteps of others, we use a broad conception of 'technology' and think that there is no good reason to see only recent inventions as technologies. But even if you disagree with the latter, there are examples of less ancient discoveries that show that not all technology is applied science, and more specifically, that the creation of new more efficient or profitable technology is not always based on scientific knowledge. A well-known example is also the symbol of the First Industrial Revolution: the steam engine.

In the history of technology, the beginning of the eighteenth century is an important period. It was then that Thomas Newcomen invented the first steam engine, which at the time was used to pump up water from mines. Nevertheless, that first model suffered from some problems, the greatest of which was low profitability. That problem was solved by James Watt, who made the first modern steam engine in 1775. Watt's solution consisted of installing an additional vessel, in which the steam was condensed. As a result, the steam engine consisted of two parts: the condenser, which was kept constantly cool, and the boiler, which had an invariably high temperature, and therefore no longer had to be heated each time, as was the case with the first steam engine. The consequences of this operation were of historic importance. The machine's performance increased by at least 20%, allowing Watt's machine to be used for heavier wagon loads, among other things.

Now the point is that no scientific knowledge was used for this improvement. Watt knew that his model was better than Newcomen's design, but he could not explain the principle underlying the fact that a steam engine with a boiler and condenser is more profitable than one without. He had arrived at his discovery through tinkering and experimentation, not by applying existing knowledge. That knowledge simply did not exist. The principle was only discovered a few decades later, more precisely in 1824. It was physicist Sadi Carnot in particular who understood how Watt's machine worked. The higher profitability had to do with the fact, Carnot discovered, that heat moves from a body with a high temperature to a body with a lower temperature. In short, the idea that you always need science for technology and innovation may be a deeply held belief, but it is not true. Of course, not all things made by humans result from ignorance, but sometimes technology development is the result of blind searching and digging, and not of knowledge, let alone scientific knowledge (Basalla, 1988).

In the Line of History

Our stories about the keyboard and the steam engine are based on the assumption that the succession and innovation of technologies are indeed unstoppable. In other words, we have assumed in the last few pages that the latest version of the determinism thesis, the one about the evolution of technology, is correct. It is now time to take a closer look at that assumption. We focus on the three best-known arguments: incremental evolution, patterns in technology development, and Moore's Law. Do these provide sufficient support for the latest version of technological determinism? That is the question on which we now focus and which, we can already reveal, we answer in the negative.

The first argument presents a historical perspective on technology, contrasting with the concept of simultaneous evolution, which includes a geographical dimension. This approach emphasizes trends in technological history, where inventions build upon and extend previous ones. Consider the evolution of ships, for example. The earliest form of a ship resembled today's kayaks or canoes: simply a log or piece of wood propelled by hand paddling. A significant development occurred when sailors began standing up, realizing that the wind could catch their clothes and propel the vessel forward, leading to the innovation of the sailboat. Over time, ships underwent continuous, incremental improvements, representing a steady stream of innovation without drastic changes from one model to the next (Basalla, 1988).

Another example is the cotton engine. This is a technology that serves to separate cotton fibers from seeds, allowing clothing to be made with those fibers. The standard version of the story states that it was invented in 1793 by Eli Whitney. On a cotton plantation in Georgia in the United States, he is said to have noticed that it was not easy to remove the seeds from short-fiber cotton, and to have realized that that cleaning process was time-consuming and very monotonous—the work was done by African American slaves (not only AI is founded on exploitation). He then observed human hands separating the fibers from the seed and translated that technique to the cotton engine. So much for the standard version. In reality, Whitney did not have to invent everything. He could rely on the Indian machine that was already in use in Italy in the twelfth century and was introduced to the USA in the early eighteenth century. Not that Whitney did not have to change anything. The Indian version was made for long-fiber cotton, whereas his machine had to be for short-fiber cotton. But Whitney was thus able to fall back on an existing model, which he modified slightly to create the version that began circulating in 1793 and greatly changed cotton cultivation in the southern USA (Basalla, 1988).

The fact that a line can be drawn between artifacts proves, according to some, that there is necessity in technology development. Assuming that is justified, it is at least not true of all things. Not every design fits seamlessly into a predecessor, not every technology lies on a continuum. The jet engine and radar, for example, are not new, slightly modified versions of existing technologies. But suppose that once they were, that all technologies are part of a step-by-step, gradual evolution. Can you then deduce that every new version of a design should exist?

It is tempting to answer in the affirmative. If you can picture history as an uninterrupted incremental upward trend, then it quickly conjures up the idea of inevitability or necessity. Further, determinism implies that technologies are an extension of each other. If it turns out that one technology inevitably leads to a new technology, then that implies that the new version follows on from the previous one, or that at least it would not be surprising if it did. However, the reverse is not true. It is possible that technologies that derive from each other necessarily do so, but the fact that technologies derive from each other is not in itself sufficient to conclude that that particular development had to occur. For example, one can imagine a government deciding to develop an AI facial recognition system that closely matches an already existing design. However, the fact that the technology resulted from a decision, a decision that also could not have been made, means that the development is not inevitable, even if there is a clear line in the development of the technology.

Patterns in Development

Perhaps the second argument for the latest version of the determinism thesis offers more support. This argument is reminiscent of the phenomenon of simultaneous evolution that we discussed earlier in this chapter. It points to patterns in the succession of old and new technologies. An invention follows another invention, and this is so, not just in one place, but in several places. Some examples: first flints were discovered, then people were able to make fire; the invention of the knife follows fire; metal making is preceded by pottery; after metal comes electricity; large-scale global communication follows the invention of electricity, and so on. You see this sequence across different cultures, different places, different countries, and this, some say, means that a chain of inventions must have occurred. The recurrence of the same sequence of technologies shows that this evolution is inevitable, so the reasoning goes.

Our response to this is similar to our response to the reasoning that refers to simultaneous evolution. The observation that the same sequence occurs in multiple places does not in itself justify the claim that that chain of designs is necessary. It cannot be ruled out that the same sequence is coincidental. But even if it is not a coincidence, that does not mean that evolution is necessary. There may be a good reason why in different places one technology comes after another, without that chain being inevitable. For example, the fact that textiles everywhere arise only after sewing is invented is because there can be no textiles without sewing. Suppose, however, that the multiple occurrences of the same sequence suffices as an argument for necessity. Even then you cannot conclude that every chain of inventions is inevitable. Some inventions that result from earlier inventions occur in different places, but this is not true of everything.

For the sake of clarity, we want to underline this. Our contention here is not that there is no necessity in technological evolution. We are simply claiming that you cannot draw that conclusion based on the observation that the same sequence of technologies returns in different places.

Moore's Law

The third and final argument for the determinist version of technology innovation is the most popular. It comes from engineer Gordon Moore, the co-founder of technology company Intel Corporation. In 1960, Moore was attending a conference when he heard colleague Doug Engelbart give a fascinating talk about chips: an assembly of electronic components such as transistors that are integrated onto a slice of silicon. Today they are used for computers, cars, cell phones, payment cards, pets, and many other things. Engelbart questioned whether scale would be beneficial in the context of electronics. He had previously observed that this is true for airplanes: the smaller they are, the better they fly. Would the same be true for technologies that work with chips? This was what Engelbart was wondering.

It was not Engelbart himself but Moore who tried to answer this question. He began keeping numerous records: the price of a transistor, the number of transistors per silicon slice, the processing speed of chips, and so on. On 19 April 1965, he presented his results in the now famous paper 'Cramming more components onto integrated circuits' in the journal *Electronics*. It showed that chips consisted of four components the first year, eight the next, and by 1965, the year of publication, more than 60, according to Moore's counting (Moore, 1965). In other words, he had found that the number of components in chips was doubling every year. Projecting those findings into the future would mean that in 1978 there would be half a million components on one chip. That turned out to be a somewhat overly optimistic prediction. Moore therefore revised his prediction in 1975: the number of transistors would double every 2 years instead of every year. That formulation has since become known as Moore's Law (Kelly, 2011).

Moore's findings are interesting for technology optimists and appeal to the imagination; and even today, there are still cases that support them. To name but a few: smartphones are becoming lighter and lighter, and the images we see on them are of increasingly better quality. In addition, we also know that on the chip A8 of the 2014 iPhone there are no less than two billion transistors on 89 mm^2; the 2013 chip A7 had only half that. Still, the question arises whether Moore's Law provides support for determinism. There are at least three problems.

First, one cannot capture all technological innovations in a law. Yes, there are other laws that are relevant in this context. Kryder's law—named after engineer Mark Kryder, former CEO of Seagate—states that the amount of data stored per square inch of a magnetic hard drive doubles every 13 months, which also lowers the price of the technology. But numerous other innovations, and perhaps even most of them, do not follow such laws. Thus, those who believe that all innovations are necessary cannot decide this on the basis of a law, because you cannot describe all innovations according to a law.

Second, there are physical limits to miniaturization. It may be true that for several decades the number of transistors doubled biennially, but that increase cannot continue indefinitely. There are limits to technological innovation, not because they should not be allowed for moral reasons, but because the innovations are physically impossible.

The third problem is more fundamental. Suppose that all technologies improve a lot. And assume for a moment that there are no limitations: the number of improvements that can be made is endless. Do we then have a strong argument in favor of determinism? In other words, if every technological innovation satisfies a law, are they all necessary?

At first glance, one must answer in the affirmative. Numerous well-known laws—the law of gravity and Boyle's law, for example—apply no matter where or when; the phenomena they describe are necessary. Yet law and necessity are not inseparable; usually, laws deal with events that happen in all cases, but not all laws deal with an inevitable course of events. An example is the law of supply and demand. It states that the price of a product depends on the relationship between supply and demand. We know that if supply is high and demand is low, the price will fall; if supply is low and demand is high, the price will rise. But, and this is crucial now, that state of affairs described by the law of supply and demand might not be there either. We can imagine a world in which it is agreed that the price of a good or service is adjusted by an algorithm, and not by the behavior of buyers and sellers. This may seem undesirable, but there is nothing to indicate that it is utterly impossible. In short, that a state of affairs in the world has a law-like character does not necessarily imply that that state of affairs also has a necessary character.

The same is true of Moore's Law. Assume for a moment that this law is true. Every 2 years the number of components on integrated circuits doubles, making technologies not only lighter and thinner, but also cheaper and more efficient. Does that mean, however, that the reality to which Moore's Law refers is inevitable, that it is impossible that they could not have been there? No, and this is why. It is not difficult to imagine a world of cyberutopians and technophiles who will do anything to make better chips so that the world will at least improve technologically. On the other hand, one can also imagine the world being run by a few members of the Amish who are not very keen on technological improvement, who believe that faster, more efficient, lighter, and cheaper technology is undesirable and will not do us much good, and who therefore seek to inhibit technological innovation. In that case, there would be no Moore's Law. Again, we do not think that it is undesirable for experts to strive for profitability and efficiency—quite the opposite—although improvement should not be reduced to those two things. But if Moore's Law is founded on the belief of a few zealots that technology must be constantly refined, it does demonstrate that what is described by Moore's Law is not inevitable.

So, what is the conclusion? Do the three arguments—incremental evolution, patterns of technology development, and Moore's Law—support the latest version of technological determinism? Anyone who concludes from the previous pages that there is no necessity in technological innovation is proceeding too hastily. We have not shown at all that determinism about the evolution of technologies is wrong and that we therefore have social control of technology development. It is still possible at this point in our argument that innovation is necessary. We have only shown that the proposal that technological innovation is necessary is not supported by the three arguments cited. A certain course of events may be described by a law, but it does not follow that that chain of events is necessary. Nor can you draw that conclusion on the basis of the recurrence of a sequence of inventions in different places and the line you might be able to draw between a succession of innovations.

Inevitable and Responsible

We have explained and evaluated technological determinism concerning the origin of technology, as well as the version concerning technological evolution. But what is the relevance of this? Why did it make sense to dwell on it for this long?

We have already touched upon this: one of the possible reasons why there is interest in technological determinism is that, according to some, it would absolve the creators of technology of their moral responsibility. If technology is inevitable, the reasoning goes, then the creators are not morally responsible for its undesirable consequences. After all, you cannot punish someone for a bad consequence of the technology you designed if it turns out that the technology was going to be there anyway, right? At least that is the thinking of some technologists and entrepreneurs. Based on the preceding pages, one cannot say that this reasoning is wrong. However, you can say that one cannot invoke technological determinism to avoid moral responsibility, simply because the arguments for determinism are flawed. If it is not established that technology is inevitable, you cannot say that you are not responsible because the technology is inevitable. Note that there are situations in which no one is responsible. A person can do something morally wrong but remain unaccountable, and thus not be a candidate for punishment. Determinism, on the other hand, cannot be invoked to erase one's moral responsibility.

Now suppose, however, that the arguments are sufficient. Moore's Law, for example, supports the determinism thesis. Does the inevitability of technology then imply that there is no moral responsibility? Can you only be morally responsible for a technology when it is not inevitable? No. To support that, we use Harry Frankfurt's famous thought experiment.

Imagine that you are an engineer and that you can create a morally undesirable technology: an AI system for job applications that excludes people who are non-native speakers of a certain language. You are not thoroughly depraved, however, and regularly doubt that you will put that system together effectively. Your neighbor, however, embodies evil and is also a surgeon. He plants a chip in your head during your sleep that can detect when you make a final decision about whether or not to make the technology. If you were to decide not to make it, the chip would take over from you and make sure that you make the technology anyway, against your will. The next morning, you wake up and decide to make the AI system. You build the system with full conviction and the brain chip did not have to interfere. Clearly, you are morally responsible for the technology. Nevertheless, the technology is inevitable. After all, if you had not created it, the chip would have; thus, the technology would have come into existence in any case. Of course, you would not be morally responsible if you had decided not to make the technology and if that chip had taken over from you. To quote the terms from Chap. 2, you would then be causally, but not morally, responsible. But that is not the point. What matters is that you decided to make the bad AI system yourself and that you knew very well what you were choosing. That makes you morally responsible, despite the fact that the technology would have been there anyway, despite the inevitability of the system.

So even assuming that an argument such as Moore's Law supports determinism, that does not absolve technology developers of moral responsibility. But what if we are wrong? What if determinism means not being morally responsible after all? Does that mean that designers and computer scientists, technologists and AI developers, can wash their hands clean of moral responsibility? If you think so, it is because you believe that technological determinism is indeed correct, despite the fact that the arguments just made fall short. But is that right? This is the question we answer in the last section. We answer it in the negative. The determinism concerning the emergence and renewal of technology is not correct, at least not for all technologies, either smart or dumb.

Technology as a Social Construct

We have arrived at the last section of this chapter. We have discussed four forms of technological determinism. The first version was about an instrumental way of thinking, the second about the social effects of technology, the third about the emergence of technology, and finally, the fourth about technological evolution. When we talked about the first two meanings of the determinism thesis, we did two things in each case. First, we showed that the support for these two versions of technological determinism is not sufficient; then, we showed that neither form of determinism is correct. In addition, when the other two versions of the determinism thesis—concerning the origin and evolution of technology—were considered, the arguments for determinism were attacked. But that does not mean that both forms of determinism are wrong. As was pointed out earlier, if it turns out that the arguments for the claim that something exists are flawed, that does not mean that it does not exist. Thus, the work is not yet complete; it remains to be shown that both the emergence and the evolution of technology are not necessary, that technology is indeed a matter of discussion, deliberation, decision-making, and compromise. To do this, we invoke two examples: algorithmic scheduling software—smart technology—and the automatic textile spinning machine—dumb technology.

Starbucks' Schedules

In 2014, journalist Jodi Kantor wrote the notorious article 'Working Anything but 9 to 5' for *The New York Times* (Kantor, 2014). In it, she tells the story of Jannette Navarro, a single mother of 22 who lives on the edge of poverty. Her employer, Starbucks, pays her only nine dollars an hour as a barista. Another problem is that her job causes her stress, a lot of stress. That is because the American company has recently started using algorithmic scheduling software. The composition of work schedules for Starbucks employees is no longer done by managers, but by an AI system bought from the company Kronos Incorporated that determines when exactly someone should work. Based on data from past consumer behavior, algorithms predict on which day the most customers will come into the coffeehouse and at what time of day the most coffee will be consumed. A high probability of passage means that many baristas are needed; if the prediction is that few customers will come, then fewer workers are needed. This is more advantageous for Starbucks itself, because such planning of work schedules means that fewer workers have to be paid at a time when they are not actually needed. But it also means that someone like Navarro can hardly plan her life a week in advance. She always has to be on standby. If the algorithm predicts on Monday that there will be a lot of customers on Saturday, Navarro has to conform to the algorithm's calculation, the power of numbers. It is an example of how AI systems can be inhuman in the double sense of the word: no human is involved in the probability calculation and the consequences are unlivable for the employees.

The introduction of AI here is clearly founded on the pursuit of efficiency, which is not exceptional. But automation can also be grounded in things other than efficiency. The goal may be repression: the desire to silence workers. This is the explanation for the switch from the hand-powered spinning machine to the automated spinning machine in the first half of the nineteenth century. Let us explain that.

Twisting loose fibers of wool or cotton together into thread with which to sew or weave is a very slow and monotonous process. From this perspective, the invention of spinning machines in the late eighteenth century was very welcome. But of course, workers were still needed to operate the machines. This was done by the spinners; their expertise was needed to make the spinning machines work. These workers were therefore in a good bargaining position when they sat around the table with the factory owners. They were able to negotiate favorable working conditions, longer breaks, and higher wages. But, crucially, this was not to the liking of the entrepreneurs. They therefore resorted to drastic measures to break the power of the spinners. After a 3-month strike, they gathered engineers and asked them to invent a new technology: the automatic spinning machine. This would have to replace the old manual machine, eliminating the need for the expertise of the spinners and preventing them from making high demands. In short, they had a new technology in mind to keep the workers down. In 1824, the factory managers got what they wanted: inventor Richards Roberts created the first automatic spinning machine. The spinners were not laid off en masse; they still provided repair and maintenance services for the machines. But their strong bargaining power collapsed, as did their willingness to strike (Basalla, 1988).

TINA, There Is No Alternative

The automatic spinning machine is an example of how an old technology is replaced by a new one; Starbucks' scheduling software is an illustration of the introduction of a new technology. Both cases show that the two forms of determinism concerning technology development are wrong. Let us recall what these brought out. All technology arises inevitably, independent of time and place; the evolution from one technology to another is necessary, and it would be there even if we could turn back time. Earlier we showed that the arguments for both interpretations are insufficient, but with the help of the two cases we now also know that neither interpretation is correct, at least for the two examples we have described. Why?

Roberts' invention and Starbucks' algorithm are clearly effects of a well-defined social relationship. The automatic spinning machine was invented at the request of factory managers to break the power of the spinners; the AI system was introduced for efficiency, to avoid paying workers when they are of no use. Both technologies would not exist outside of that context; if that context were not there, neither would have been developed. The point now is that both social relations are also the product of industrial society, which itself does not follow a law of nature, and which, in turn, is not necessary. Not every society is grafted onto the distinction between, on the one hand, capital and free entrepreneurs, and, on the other, labor and workers. Moreover, in the light of history, industrialization is fairly recent. It has not always existed; it emerged from the beginning of the eighteenth century.

Our thesis therefore reads as follows. The spinning machine and planning software are not technologies that needed to be created. We are not commenting here on their desirability, but a world without both was and is not utterly implausible. If we could turn back time, you could not be sure that they would come to exist again. This conclusion is based on the following reasoning. In a society not centered around labor and capital, these technologies would not have been invented. And because such a society is not inevitable, neither are the spinning machine and Starbucks' AI system. This is the case for these technologies, and probably for many others—maybe even for all technologies. You cannot rule out the latter, just as it is not impossible that some may indeed be inevitable.

Thus, the version of determinism concerning the origin and development of technology is not correct. Note that we are *not* claiming here that *no* technology is inevitable, rather that not all technologies are. Some technologies do not fit the framework of determinism, which of course is a reason to suspect that there are other examples that do not fit the determinist's worldview. At least, that is what the cases of the automatic spinning machine and Starbucks' scheduling software make clear. But you can also add the Big Five to that. Take Google. The founders of the company, Page and Brin, instead of creating the tech giant, could have become professors at their alma mater. They could have chosen not to commercialize the search engine. And it was not inevitable that they opted for a business model that made them rich, a model based upon targeted advertising. At least we might not have lost so much privacy over the last decade.

Suppose, however, that our reasoning errs. It was inevitable that the industrial society came into being and that it was based on the distinction between employer and employee. Would that then also signal the end of our attack on determinism? Does it follow that the technologies in our two examples were inevitable?

The answer to that is in the negative. Although it may be true that the spinning machine and Starbucks' AI would not exist outside of industrial society, the relationship between those technologies and industrialization is not the same as the relationship described by Boyle's gas law. The spinning machine and planning software do not result from industrial society in the same way that the increase in air pressure follows from the decrease in the volume of air in a bicycle pump. It is inevitable that air pressure will rise, but it is not sure that the spinning machine or the algorithm will be used. This is because the technology is the result of a choice, a decision made by one or more people in a particular social role: the employer, management, and so on. This was evident in the case of the spinning machine, but it is also true in the Starbucks case. When Howard Schultz, the company's former CEO, heard about the stories brought out by *The New York Times*, he apologized and said that the creation of work schedules was open to revision. Certainly, the decision to use an automatic spinning machine or scheduling software is founded on reasoning: to suppress protests and to seek efficiency, for example. But a reasoned decision is still a decision, and that means that there were multiple options and that, ultimately, things could have gone differently.

The SCOT Approach

Let us be clear about the background of the story that we have outlined on the previous pages. This story ties in with a movement from the second half of the last century known as social constructivism, which is associated with the work of Michel Foucault, among others. More specifically, our argument falls within the so-called SCOT approach, or the Social Construction of Technology. One of the best-known representatives of that approach is the aforementioned Bijker, whose social constructivist view of the development of the bicycle has been very influential.

Social constructivism does not offer a detached, birds-eye view of technology. On the contrary, it functions at a concrete level and examines the genesis of technologies there. One uncovers the social factors that led to the existence of what is claimed to be a social construct. That is, the interpersonal relationships, groups, and social roles that have shaped the development of technology are detected. The examples of Starbucks and the automatic spinning machine fit within that context, just as Bijker's research shows that differences in preferences among people have led to different types of bicycles.

Other examples include the studies that uncovered funding streams for AI (Eynon & Young, 2021). Specifically, it is well known that many technologies that are not AI have military origins—the Internet, aviator sunglasses, sanitary pads—but we also know that the military contributed to the heyday of AI. The event often seen as the beginning of AI—the famous 1956 Summer Research Project on Artificial Intelligence at Dartmouth College in New Hampshire—was funded by the US Department of Defense Office of Naval Research. It also includes the Defense Advanced Research Projects Agency (DARPA), which we cited earlier in Chap. 1. Its main mission is to manage research funds. Robert Sproull, former director of the institute, once said that at least two generations of computer scientists were funded by DARPA, and so, among other things, technology that understands language has been financially supported by the Department of Defense. The same institute has also contributed to the development of self-driving cars. In 2004, the first DARPA Grand Challenge was organized. This was a competition for cars to drive autonomously for over 200 km in the desert between Nevada and California. The winning team was rewarded with a hefty sum of money—the first year it was US$1 million, and the following year it was double that amount. Finally, it is perhaps no coincidence that Amazon had one of its newest headquarters built near the Pentagon.

Looking at technology in a social constructivist way, then, means focusing on the social cause of technology, rather than the social consequences. The point is not to determine on which social roles and relationships technology exerts a good or bad effect. It is the other way around. Within such a framework, the point is to ask which social roles and relationships underlie technology. For whom is technology made? In what context is it embedded? Who decides? What interests played a role? Yet this approach can also be linked to the perspective from which to focus on the consequences of technology in the following way.

Suppose a company uses an AI system that constantly monitors employees with the intention of knowing whether they are working efficiently. This creates a problem: quite a few employees are very uncomfortable with it, to the point that some truly suffer from the continuous surveillance. Well, for some, these negative consequences of the use of the surveillance system on the workers are precisely the reason to be interested in social constructivism. On the basis of the undesirable influence, they want to trace the genesis of the technology, and find out which parties were involved in the choice of that technology. There may be a psychological explanation for this. To channel dissatisfaction with a problem, it helps for some people to know the origin of that problem. But there may also be moral motives at play. One looks to social history, hoping to find a person responsible for the problem at hand.

The link between the perspective that focuses on social consequences and a social constructivist perspective can also have a militant character. Let us look at the control system from earlier. Not only is that technology annoying, it is also having undesirable effects on some workers. The pressure to work as hard and efficiently as possible is so great that there is a danger that some will no longer be able to function and will have to stay at home for long periods of time. For some, therefore, it is clear: we would be better off without it. Because of the bad effects, they want to return to a work context without AI and surveillance. The question that then arises is whether this is possible. A company without the technology is desirable, but is it possible? Some take a social constructivist view of technology with this question in mind. They then reason as follows. When you get a good view of the social context within which the system was developed, then you also know that it was not inevitable that the technology was created at that time. And if you know that the technology could not have been there either, then things can be different now. At least, that is the underlying reasoning, which is founded on the assumption that there is a link between being avoidable and being changeable. It is assumed that something can be different now because it could have been different then. We elaborate on that assumption later.

Between Dream and Deed

On the opening page of his study *The Social Construction of What?* philosopher Ian Hacking notes that in recent decades a great many things have been said to be social constructs: gender, emotions, science, illness, refugees, and a host of other things (Hacking, 1999). Even facts are called social constructs by some, and as recently as 2020 the famous philosopher Giorgio Agamben suggested that the corona epidemic might be an invention—read: a social construction—of (bio)politics. We highly doubt that all of these things are equally plausible, but we would nevertheless like to add Starbucks' automatic spinning machine and scheduling software to this list. Both technologies are social constructs and so they might as well not have been there. In that respect they correspond to legal or political laws but differ from a phenomenon such as the increase in air pressure when the volume of air in a bicycle pump decreases.

To avoid ambiguity, let us emphasize this: perhaps many technologies are social constructions, but we do not claim that this is certainly true of *all* technologies. Our claim is that some are socially constructed; thus, not every technology is determined. That claim may be crystal clear, but we would like to draw attention to five possible misunderstandings. The first has to do with the difference between can and will, the last with the link between determinism and constructivism.

1. Earlier, we pointed out that there are other forms of determinism besides the technological variant, including the genetic variant. We now add social determinism. It means that the desires associated with social roles, processes, and relationships will be satisfied in all cases. When talking about this type of determinism in the context of technology, it is clear that it is always accompanied by social constructivism. If the desire of a group of people for a particular bicycle to be made is fulfilled anyway, then that bicycle is a social construct. However, the reverse is not true. Social constructivists can be social determinists, but that is not necessarily the case. You can believe that technologies are social constructs because they are the outcome of the will of a number of people in a well-defined social relationship. But at the same time, you can realize that the desire for a particular technology to be made cannot be fulfilled, for example, because of physical limitations or because it is organizationally impossible. There is another way of saying this: there is a gap between having a wish and being able to fulfill that wish. Many technologies that can be developed are not wanted by many people. We are thinking primarily of the atomic bomb or the killer robots from Chap. 2, but there are also less dramatic

examples. An app that gives developers access to the data of all your contacts: it is possible, but most of us find it undesirable, and rightly so. Conversely, many people may wish for a technology when in fact it cannot be made. Consider Moore's Law. A number of technophiles may desire that even in 2030 the number of transistors will continue to increase dramatically, but it will not happen because of the physical limit upon which designers irrevocably stumble. Not everything that is possible is desirable, but not everything that is desirable is possible either.

2. The claim that a technology inevitably follows from an old one certainly does not apply to all things. Between the old and the new spinning machines there was a social relationship in a specific era that might not have necessarily been there, and so the automated version is not inevitable either—we saw that earlier. But what if that were now true of all technologies? Should we then conclude that earlier technologies play no role, and that the development of new designs is always independent of existing designs? Think of the early days of the PC. Back then, it was controlled by the MS-DOS program, which had a working memory of 640 kb. Although that capacity was large at the time, it soon proved to be limited. Nevertheless, it was not until much later that an entirely new system was developed: Windows 95 by Microsoft. Initially, they had developed a program that was based on the original model but had an expanded working memory. This ties in with what we saw earlier about the ship and the spinning machine. For the development of many technologies (but not all), one starts from an existing technology and then makes a new more efficient version that is not completely different after all. This usually has to do with economic motives. If money is pumped into a technology and then one were to design a more efficient but entirely new technology, it would negate those investments. To avoid that, the existing model is taken as the starting point for the new and more efficient version. So, even if all technologies are social constructs, it does not necessarily mean that technology development is exclusively the result of a desire of a few people and of what is physically or organizationally possible. Technology development in the present can also be determined by technology development in the past. Although we would like to repeat that it is not at all certain that the first technology will inevitably result in the second. A more plausible claim is that an existing technology makes the development of a new version not necessary but likely, and that this has to do with the financial investment in the first technology.

3. The concepts of real and constructed are often contrasted. What is constructed is not real; only what is not constructed is real. If a technology is a social construct, then the implication is that it is not real. This, of course, is not true. Starbucks' scheduling software, for example, is the outcome of a social relationship, but it is embedded in hardware of a tangible and measurable nature. Moreover, despite the fact that the technology is a social construct, the effects of the AI system are strongly felt. At least, that is what emerges from the stories that Kantor chronicled in *The New York Times*, stories woven around stress, uncertainty, and fatigue. Or consider the examples that recurred in the various chapters: the unequal treatment of people by the biased smart systems used by courts, police, employers, and so on. These are

undesirable effects of systems that are social constructs. But that does not make the effects any less real.

4. We have just cited the spinning machine and planning software to show that the last two versions of technological determinism—those about the emergence and renewal of technology—are wrong. Because we showed afterward that both technologies are social constructions, the impression might now arise that technological determinism and social constructivism are mutually exclusive. This is, of course, correct when it comes to the latter two variants of determinism. If a technology is a social construct, then it might as well not have existed, which is exactly the opposite of what the determinist claims. Nevertheless, the two can also go together. We refer then to the second form of determinism, which is that a technology inevitably has social effects. Certainly not all technologies have necessary social effects—we showed that earlier—but let us assume that at least one technology does. If you focus on that, it turns out that such a design could be the result of a social process, and consequently that it might not have been there. In other words, it is not excluded that a technology has unavoidable consequences on a social level, while at the same time being the result of a well-defined social relation at a certain point in time, and thus is not itself unavoidable.

5. A technology can thus be both a social construct and determine something else, but a technology can also be socially constructed and determined, albeit not in the sense that the technology should have absolutely been created. To illustrate this, we turn to a technology that, like the Internet and the computer, has military roots and is used today in at least 30 countries to generate electricity: the nuclear power plant. There is no doubt that nuclear power plants, unlike, say, natural phenomena, are social constructs. They are the result of a decision made by at least three stakeholders—politicians, industrialists, and scientists—and they consist of a choice that also might not have been made. So the fact that nuclear power plants are used is not inevitable; it could have been different. But, importantly, at the time the decision was made to develop this technology, in part, the nuclear power plant cannot be different from how it is. We are talking here about what makes a nuclear power plant a nuclear power plant: the generation of energy through the fission of uranium that is converted into electricity. In other words, that the technology exists is not necessary, but what makes it a nuclear power plant is inevitable—at least in part. This is because, unlike the steam engine, the development of the nuclear power plant is based on science, and more precisely, on knowledge of nuclear physics. And that knowledge, in turn, is dictated by how reality works at the physical level. Of course, in a different world with different funding channels, the field of nuclear physics might not have emerged. As a discipline, it is a social construct, but the knowledge that it produces is not. Scientific knowledge of physicists follows the structure of reality and consequently cannot be other than it is. And that obviously has consequences for the technology that is founded on that knowledge and of which it is therefore incorrect to claim that it was not inevitable. A technology can therefore be both determined and socially constructed. Anyone who thinks that this is a blatant contradiction should bear the following in mind. That a technology is constructed means that it results from a decision made by a group of people. The constructed

character thus refers to its genesis. When we claim here that a technology is determined, we are talking about the technology itself, not its genesis. The structure of a nuclear power plant cannot be other than it is, and that follows from the construction of reality, a construction that the iron law of nuclear physics imposes on technology.

Constructivism and Activism

Clearly, there is something at stake in our account of social constructivism. We have introduced this approach because it at least challenges and even undermines the two latest versions of the determinism thesis. In addition, our story is relevant in a practical sense. That is because the determinism thesis often pops up in discussions, political or otherwise.

Take the killer robots from Chap. 2, a technology that is regularly debated up to the level of the United Nations. An oft-repeated argument by proponents is that they will come anyway. The reasoning is that killer robots will in all cases be used in the future, and so you might as well allow them now. If the international community were to go along with this, it could potentially have serious real-world consequences. These include the possibility of a humanitarian disaster—robots might make mistakes and kill countless civilians—and the risk of lowering the threshold for going to war, because the use of such technology reduces the number of military personnel who need to take up arms effectively. Now, however, if you can show by means of the automatic spinning machine that technological determinism is wrong in a general sense, then you have a reason to at least question the argument of the proponent of killer robots. If such technology is not deterministic, why should hyper-advanced AI systems be? That question forces the proponent to substantiate his reasoning, which may be to the proponent's disadvantage. This may be of little or no relevance when it comes to technologies without (much) impact, but it is different when it comes, for example, to fully automated weapons systems that can claim human lives.

Some add the following. A social constructivist perspective is useful because it implies that the current situation can also be changed. If a technology that exists today also could not have existed, then a society without that technology today is possible, so the reasoning goes. Is that right?

It is clear that this interest in social constructivism comes from a particular angle. It is usually about people who have a negative view of a technology and who want it to disappear precisely for that reason. That negative judgment can be related to several things, things that appeared regularly throughout the chapters of this book. We have seen that both the development and the use of AI systems have undesirable ecological effects. Smart technology training, for example, involves the emission of at least 200,000 kg of CO_2. People can also be critical of a technology because of its existing or potential political abuse. Take the example we cited in Chap. 1. In the run-up to the 2016 US presidential election, false information about Hillary Clinton, Muslims, and Mexican immigrants was spread through anonymous accounts on

Twitter and Facebook. Those accounts were owned by the Internet Research Agency company, better known as the Trolls from Olgino, which is based in St. Petersburg and had the goal of influencing the outcome of the election. Another possible source of negative opinion has to do with real or potentially undesirable consequences on a personal level. There are many examples of this: quite a bit of technology enacts an outright invasion of our privacy; the use of social media can lead to symptoms of depression; social media can be used for cyberbullying, and we know from research that this can lead to anxiety, suicidal thoughts, and even suicide.

These problems are too serious to ignore, and concerns about them are justified, yet the reasoning is not without problems. Those who believe that it is better for a technology to disappear because it is accompanied by negative consequences are actually arguing for a world without (much) technology. After all, there are few, if any, technologies that are completely risk free. Everything or almost everything we make carries dangers, and so a risk in itself is insufficient to argue for the disappearance of the technology. Moreover, it does not take into account the positive effects. Of course, we know that some things are made with completely malicious goals in mind, and of course we realize that there is disagreement about whether the effects of some technologies are desirable or not. At the same time, you cannot deny that there are many inventions whose value virtually everyone agrees upon. We are thinking primarily of technology to detect or cure diseases. To argue for the disappearance of a technology solely on the basis of its negative consequences is therefore too short, not only because you would actually have to condemn many things, but also because of the positive consequences. A trade-off between the various consequences is needed, and not a one-sided view only of the undesirable effects.

We assume for a moment that the trade-off has been made and that one decides that it would be better not to have an existing technology. Can a constructivist view then support the activist's pursuit of a world without that technology? Can one infer from the understanding that a technology that exists now, but might just as easily not have existed, that a society without that technology is possible?

It is understandable if you answer that question in the affirmative. Numerous things that are the case now, but equally might not have been, can be erased today. If one's car's lights were installed poorly by the garage owner because she or he was in a hurry, that was an avoidable moment. But that mistake can be corrected—the garage owner can now install the lights correctly. Another example: the fact that nuclear power plants are used today is the result of a choice made in the last century that might not have necessarily been made then either, but which can still be revisited today. Yet we want to emphasize that one does not necessarily follow from the other. It is not because something could have turned out differently that you can undo it now. Changeable does not necessarily follow from avoidable. Think of terror. Someone with bad intentions can commit a deadly attack. The deaths of the victims were avoidable, yet their deaths are irrevocable.

The same is true of technology. Recent developments in robotics and AI make it possible to design a technology that, once put into use, cannot be undone. That design may follow from a decision and therefore could just as well not have been there, but now that the technology is there anyway, it is irrevocably so. In short, if you believe that a world without a particular technology would be a better world, and if, moreover, you know that that technology is a social construct and therefore not inevitable, that knowledge is not in itself a reason for optimism. It is not because an existing technology might as well not have been there that a world without that technology is now possible. It may be that the use of a technology is not inevitable and can still be reversed, but it is not the case that something equally could not have existed and can therefore now be erased.

The activist who opposes a technology might respond that this is only true of hyper-advanced AI systems, and not all smart technologies, let alone technology in general. The majority of existing technologies are controlled by corporations and governments, and that is a reason for hope. After all, we, activists, can put pressure on those agencies, can inform and mobilize citizens, potentially pulling the plug on a technology. Although that comment may be justified, we would like to emphasize the following two things.

First, the fact that a technology could not have existed does not mean that the existence of that technology can now be reversed—we saw that just now—but the fact that the use of a technology can now be stopped does not mean that people will actually stop using it. After all, there can be all sorts of reasons to ensure that something will not change, whereas it can hold the possibility of change. Think about cars. In the light of history, cars are a young technology, a development that might not necessarily have come into existence and about which one can decide that its use should not be continued. Our point now is not that the latter is desirable, but rather that it is by no means impossible that in the future it will be decided that cars should disappear from the streets. At the same time, it is clear that such a decision would, rightly or wrongly, face a lot of resistance, for example, because of the economic interests of car manufacturers, and also on the basis of the fact that, for several decades, governments have been committed to a society that includes cars by building streets and highways. If these concerns are shared by key stakeholders, then there is a real chance that cars will not disappear, even if they could disappear.

Second, when the decision is made to actually eliminate a technology that can be eliminated, it does not mean that there is a direct line between will and reality. Actualizing the possibility of change can be extremely slow and arduous. Again, let us look at cars. If one were to decide that cars should disappear, this would obviously not happen overnight. This slowness may be due to a psychological mechanism such as habit formation or the tendency to want to keep everything as it is. However, it could be related to other factors too. There is also the fact that the comings and goings of a significant number of people and companies currently depend on the use of cars, and that our current mode of living together is strongly intertwined with this particular transportation technology. Putting an end to the use of this technology would thus have profound consequences on multiple levels; that alone is a reason to suspect that the activist's desire for change will not be realized as quickly as desired.

In summary, we would like to underscore this. Of course, a social constructivist view of technology is relevant, not only for its attack on determinism, but also on a practical level. At the same time, there are reasons not to overestimate the practical importance of that approach. Of course, we do not mean to say that change is undesirable and that change, even if desirable, will not happen anyway. We wanted to draw attention in the last few paragraphs to issues that may slow down or even prevent meaningful change in the field of technology, even if it is entirely possible. This should be interpreted neither as an expression of conservatism nor as an aversion to activism but rather as an expression of realism.

Conclusion

Of course, not all engineers and entrepreneurs speak in terms of determinism when talking about technology and AI, but it is true for no small number of engineers and entrepreneurs. It is important to note that in this context one can distinguish four types of assertions. First, there is the theory, championed by Heidegger and others, that today we can only look at the world in an instrumental sense. The second version of the determinism thesis states that technology inevitably has social effects. The third and fourth versions deal with the history of technology, that is, with the emergence and evolution of technology. In this chapter, we have highlighted these four propositions and explained them as clearly as possible. In addition, we have taken an evaluative look at those four versions of technological determinism, with two questions in mind each time. Are the arguments given in support of technological determinism sufficient? If not, are there reasons to reject the thesis? This second question is not trivial. After all, if it turns out that an argument is not doing the job it is supposed to do, then you cannot infer that the proposition is false, as there may be other decisive arguments. At this point, when it comes to technological determinism, our conclusion is this. For none of the four versions do the arguments provided suffice. Indeed, each version can be undercut by at least one argument. Applied to the determinism thesis about the emergence of technology: simultaneous evolution does not show that the emergence of a technology is inevitable, and, moreover, Starbucks' scheduling software might as well not have been there. Now does the latter mean that no technology is inevitable? No. Not every technology is inevitable is what we are claiming by reference to Starbucks' software, and that is not equivalent to claiming that no technology is inevitable.

Afterword

As we conclude our journey through the philosophical landscape of technology, it is important to reflect on the implications of what has been discussed in this book. Throughout these pages, we sought to dismantle prevalent myths surrounding technology—its supposed neutrality, its inherently disruptive nature, and its deterministic evolution. By challenging these notions, we encourage a more thoughtful engagement with technology, urging readers to consider not just the capabilities of technological innovations but their broader ethical and societal impacts.

This book has been aimed not only at critiquing but also enlightening, offering a pathway toward a more responsible and informed approach to technology. We have seen that technologies are not mere tools but are embedded with values that reflect the biases and intentions of their creators. Recognizing this should empower us as users and developers to strive for technology that respects the values we (should) care about. It is not enough to accept technology as it is presented; we must retain a questioning attitude, examining the alignment of technologies with our moral goals.

In debunking the myth of technological determinism, this book centers on human agency as a vital factor in technological advancement. By framing technology as a product of human choices and cultural contexts, it moves away from the view of technology as an inevitable force. Instead, technology emerges as a flexible and malleable tool, shaped by the intentions and decisions of those who create and use it. This shift highlights the importance of understanding technology, not as a predetermined path but as a series of developments influenced by diverse human factors. Such an understanding encourages a broader appreciation of how technologies can be shaped to fit the varied fabric of human needs and aspirations, independent of any predetermined course.

In closing, let this book serve as a starting point for ongoing dialogue and action. The conversation about the philosophy of technology is far from over; it is an evolving discourse that must adapt to the changing contours of our technological landscape. As we continue to innovate and navigate the complexities of the digital age, let us remain vigilant and thoughtful, ensuring that our technological future is not just something that we inherit but something we consciously and ethically create.

Bibliography

Agar, N. (2015). *The sceptical optimist. Why technology isn't the answer to everything*. Oxford University Press.
Agrawal, A., Gans, J., & Goldfarb, A. (2018). *Prediction machines. The simple economics of artificial intelligence*. Harvard Business Review Press.
Algar, C. (2020, April 21). *New 10-year low in global executions, but progress marred by spikes in a few countries*. https://www.amnesty.org/en/latest/news/2020/04/op-ed-new-10-year-low-in-global-executions-but-progress-marred-by-spikes-in-a-few-countries/
Amoore, L. (2020). *Cloud ethics. Algorithms and the attributes of ourselves and others*. Duke University Press.
Aral, S. (2020). *The hype machine. How social media disrupts our elections, our economy, and our health—and how we must adapt*. HarperCollins Publishers.
Arthur, W. B. (2011). *The nature of technology. What it is and how it evolves*. Free Press.
Asimov, I. (2004). *I, robot* (Vol. 1). Spectra.
Bartoletti, I. (2020). *An artificial revolution. On power, politics and AI*. The Indigo Press.
Basalla, G. (1988). *The evolution of technology*. Cambridge University Press.
Belkhit, L., & Elmligi, A. (2018). Assessing ICT global emissions footprint: Trends to 2040 and recommendations. *Journal of Cleaner Production, 177*, 448–463.
Bijker, W. E. (1995). *Of bicycles, Bakelites, and bulbs: Toward a theory of sociotechnical change*. MIT Press.
Blenner, S. R., et al. (2016). Privacy policies of android diabetes apps and sharing of health information. *Journal of American Medicine, 315*(10), 1051–1052.
Boddington, P. (2017). *Towards a code of ethics for artificial intelligence*. Springer.
Bostrom, N. (2014). *Superintelligence: Paths, dangers, strategies*. Oxford University Press.
Brey, P. (2015). Design for the value of human well-being. In J. van den Hoven, P. Vermaas, & I. van de Poel (Eds.), *Handbook of ethics, values, and technological design* (pp. 1–14). Springer.
Brin, S., & Page, L. (1998). The anatomy of a large-scale hypertextual web search engine. *Computer Networks and ISDN Systems, 30*(1), 107–117.
Buchanan, A. (2011). *Beyond humanity?* Oxford University Press.
Bunge, M. (1966). Technology as applied science. *Technology and Culture, 7*(3), 329–347.
Butler, S. (1863). Darwin among the machines. In J. H. Jones & A. T. Bartholomew (Eds.), (1923). *A first year in canterbury settlement and other early essays* (pp. 179–180). Shrewsbury Edition of the Works of Samuel Butler. Cape.

Christiaens, T. (2022). *Digital working lives: Worker autonomy and the gig economy*. Rowman & Littlefield.

Chugh, D. (2018). *The person you mean to be. How good people fight bias*. HarperCollins.

Clark, S. (2014, December 2). Artificial intelligence could spell end of human race—Stephen Hawking. *The Guardian*. https://www.theguardian.com/science/2014/dec/02/stephen-hawking-intel-communication-system-astrophysicist-software-predictive-text-type

Claussen, J., Peukert, C. & Sen, A. (2019). *The editor vs. the algorithm: Returns to data and externalities in online news*. CESifo Working Paper Series.

Coeckelbergh, M. (2020). *AI ethics*. MIT Press.

Couldry, N., & Mejias, U. A. (2019). *The costs of connection. How data is colonizing human life and appropriating it for capitalism*. Stanford University Press.

Cramer, M. (2021, June 3). A.I. drone may have acted on its own in attacking fighters, U.N. says. *The New York Times*. https://www.nytimes.com/2021/06/03/world/africa/libya-drone.html

Crawford, K. (2021). *Atlas of AI. Power, politics, and the planetary costs of artificial intelligence*. Yale University Press.

Danaher, J. (2019). *Automation and Utopia. Human flourishing in a world without work*. Harvard University Press.

Dastin, J. (2018, October 11). Amazon scraps secret AI recruiting tool that showed bias against women. *Reuters*. https://www.reuters.com/article/us-amazon-com-jobs-automation-insight-idUSKCN1MK08G

De Ketelaere, G. M. (2020). *Mens versus machine. Artificiële intelligentie ontrafeld*. Pelckmans.

Dignum, V. (2019). *Responsible artificial intelligence. How to develop and use AI in a responsible way*. Springer.

Engler. (2019, October 31). *For some employment algorithms, disability discrimination by default*. https://www.brookings.edu/articles/for-some-employment-algorithms-disability-discrimination-by-default/

Eynon, R., & Young, E. (2021). Methodology, legend, and rhetoric: The constructions of AI by academia, and policy groups for lifelong learning. *Science, Technology & Human Values, 46*(1), 166–191.

Feenberg, A. (1999). *Questioning technology*. Routledge.

Fischer, J. M., & Ravizza, M. (1998). *Responsibility and control. A theory of moral responsibility*. Cambridge University Press.

Fisman, R. (2013, March 11). Did eBay just prove that paid search ads don't work? *Harvard Business Review*. https://hbr.org/2013/03/did-ebay-just-prove-that-paid

Ford, M. (2015). *Rise of the robots*. Basic books.

Foroohar, R. (2021). *Don't be evil: The case against big tech*. Crown Currency.

Frey, C. B., & Osborne, M. A. (2017). The future of employment: How susceptible are jobs to computerisation? *Technological Forecasting and Social Change, 114*(issue C), 254–280.

Fry, H. (2018). *Hello world: How to be human in the age of the machine*. Random House.

Gabriels, K. (2020). *Conscientious AI: Machines learning morals*. VUB Press.

Garber, M. (2013, September 20). Funerals for fallen robots. *The Atlantic*. https://www.theatlantic.com/technology/archive/2013/09/funerals-for-fallen-robots/279861/

Gray, M. L., & Suri, S. (2019). *Ghost work. How to stop Silicon Valley from building a new global underclass*. Houghton Mifflin Harcourt.

Gunkel, D. J. (2018). *Robot rights*. The MIT Press.

Hacking, I. (1999). *The social construction of what?* Harvard University Press.

Hansson, S. O. (2013). *The ethics of risk. Ethical analysis in an uncertain world*. Palgrave Macmillan.

Heaven, W. D. (2020, July 17). Predictive policing algorithms are racist. They need to be dismantled. *MIT Technology Review*. https://www.technologyreview.com/2020/07/17/1005396/predictive-policing-algorithms-racist-dismantled-machine-learning-bias-criminal-justice/

Heidegger, M. (1977). *The question concerning technology*.

Heilbroner, R. L. (1994). Do machines make history? In M. R. Smith & L. Marx (Eds.), *Does technology drive history? The dilemma of technological determinism*. The MIT Press.
Jonas, H. (1979). *Das Prinzip Verantwortung: Versuch einer Ethik für die technologische Zivilisation*. Suhrkamp Taschenbuch.
Kaiser, B. (2019). *Targeted*. Harper Collins Publishers.
Kantor, J. (2014, August 13). Working anything but 9 to 5. *The New York Times*. https://www.nytimes.com/interactive/2014/08/13/us/starbucks-workers-scheduling-hours.html?mtrref=www.google.be&gwh=0A6989B77130BE779940400A811F5F43&gwt=pay&asetType=PAYWALL
Kapp, E. (1877). *Grundlinien einer Philosophie der Technik. Zur Entstehungsgeschichte der Cultur aus neuen Gesichtspunkten*. Georg Westermann.
Kearns, M., & Roth, A. (2020). *The ethical algorithm. The science of socially aware algorithm design*. Oxford University Press.
Kelly, K. (2011). *What technology wants*. Penguin.
Lambert, D. (2019). *La robotique et l'intelligence artificielle*. Éditions jésuites.
Lambrecht, A., & Tucker, C. (2019). Algorithmic bias? An empirical study of apparent gender-based discrimination in the display of STEM career ads. *Management Science, 65*(7), 2966–2981.
Latour, B. (2012). *We have never been modern*. Harvard University Press.
Lauwaert, L. (2021). *Wij, robots: een filosofische blik op technologie en artificiële intelligentie*. Lannoo Meulenhoff.
Leveringhaus, A. (2016). *Ethics and autonomous weapons*. Palgrave Macmillan.
Levy, R. (2021). Social media, news consumption, and polarization: Evidence from a field experiment. *American Economic Review, 111*(3), 831–870.
Marx, K. (1970). *Das Kapital* (La Capital). Newcomb Livraria Press.
Matthias, A. (2004). The responsibility gap: Ascribing responsibility for the actions of learning automata. *Ethics and Information Technology, 6*, 175–183.
Miller, B. (2021). Is technology value-neutral? *Science, Technology & Human Values, 46*(1), 53–80.
Moody, O. (2020, August 13). German AI posts fake child-abuse videos online to catch abusers. *The Times*. https://www.thetimes.co.uk/article/german-ai-posts-fake-child-abuse-videos-online-to-catch-abusers-rq0hc0wxs
Moore, G. E. (1965). Cramming more components onto integrated circuits. *Electronics, 34*(8), 114–117.
Norden, E. (1969). The playboy interview: Marshall McLuhan. A candid conversation with the high priest of popcult and metaphysician of media. *Playboy magazine, 16*(3), 53–74.
O'Neil, C. (2016). *Weapons of math destruction. How big data increased inequality and threatens democracy*. Crown.
Perez, C. C. (2019). *Invisible women: Data bias in a world designed for men*. Abrams.
Petzet, H. W. (1993). *Encounters and dialogues with Martin Heidegger: 1929–1976* (P. Emad & K. Maly, Trans.). University of Chicago Press.
Pitt, J. (1987). The autonomy of technology. In P. T. Durbin (Ed.), *Technology and responsibility. Philosophy and technology* (pp. 99–114). Springer.
Pitt, J. (1999). *Technological determinism. Foundations of the philosophy of technology*. Seven Bridges Press.
Pitt, J. C. (2014). "Guns don't kill, people kill"; Values in and/or around technologies. In P. Kroes & P.-P. Verbeek (Eds.), *The moral status of technical artefacts* (pp. 89–101). Springer.
Rosen, L. D., Cheever, N. A., & Carrier, L. M. (2015). *The Wiley handbook of psychology, technology, and society*. Wiley Blackwell.
Russell, S. (2019). *Human compatible. Artificial intelligence and the problem of control*. Viking.
Savulescu, J., & Maslen, H. (2015). Moral enhancement and artificial intelligence: Moral AI? In J. Romportl et al. (Eds.), *Beyond artificial intelligence* (pp. 79–95). Springer.
Schor, J. B. (2020). *After the gig. How the sharing economy got hijacked and how to win it back*. University of California Press.

Schwarz, E. (2018). *Death machines. The ethics of violent technologies.* Manchester University Press.
Smith, M. R. (1994). Technological determinism in American culture. In M. R. Smith & L. Marx (Eds.), *Does technology drive history? The dilemma of technological determinism* (pp. 2–35). The MIT Press.
Soper, S. (2021, June 28). Fired by bot at Amazon: 'It's you against the machine'. *Bloomberg.* https://www.bloomberg.com/news/features/2021-06-28/fired-by-bot-amazon-turns-to-machine-managers-and-workers-are-losing-out
Srnicek, N. (2017). *Platform capitalism.* Polity Press.
Stiegler, B. (2019). *The age of disruption. Technology and madness in computational capitalism.* Polity Press.
Tegmark, M. (2018). *Life 3.0: Being human in the age of artificial intelligence.* Vintage.
Trafton, A. (2020, February 20). Artificial intelligence yields new antibiotic. *MIT News.* https://news.mit.edu/2020/artificial-intelligence-identifies-new-antibiotic-0220
Vaidhyanathan, S. (2018). *Antisocial media. How Facebook disconnects us and undermines democracy.* Oxford University Press.
Van de Poel, I. (2015). Values in engineering and technology. In W. J. Gonzalez (Ed.), *New perspectives on technology, values, and ethics* (pp. 29–45). Springer.
Van de Poel, I., & Kroes, P. (2014). Can technology embody values? In P. Kroes & P.-P. Verbeek (Eds.), *The moral status of technical artefacts* (pp. 103–124). Springer.
Van de Poel, I., & Royakkers, L. (2023). *Ethics, technology, and engineering: An introduction.* John Wiley & Sons.
Van de Poel, I., Fahlquist, J. N., Doorn, N., Zwart, S., & Royakker, L. (2012). The problem of many hands: Climate change as an example. *Science and Engineering Ethics, 18*, 49–67.
Veletsianos, G. (2014, January 23). *On Noam Chomsky and technology's neutrality.* https://www.veletsianos.com/2014/01/23/on-noam-chomsky-and-technologys-neutrality/
Véliz, C. (2021). *Privacy is power. Why and how you should take back control of your data.* Melvillehouse.
Wallach, W., & Collin, A. (2009). *Moral machines. Teaching robots right from wrong.* Oxford University Press.
Warnier, M., Dechesne, F., & Brazier, F. (2015). Design for the value of privacy. In J. van den Hoven, P. Vermaas, & I. van de Poel (Eds.), *Handbook of ethics, values, and technological design* (pp. 1–14). Springer.
Whittaker, M., et al. (2019). *Disability, bias, and AI.* https://ainowinstitute.org/disabilitybiasai-2019.pdf
Winner, L. (1980). Do artifacts have politics? *Daedalus, 109*(1), 121–136.
Zuboff, S. (2019). *The age of surveillance capitalism. The fight for a human future at the new frontier of power.* Profile Books.

Recommended Literature

A few resources are grouped by chapter below, which may help those interested to explore the topic further.

Chapter 1: The Neutrality of Technology

Klenk, M. (2021). How do technological artefacts embody moral values? *Philosophy and Technology, 34*, 525–544.
Kroes, P. A., & van de Poel, I. R. (2014). Can technology embody values? In P. Kroes & P. P. Verbeek (Eds.), *The moral status of technical artefacts* (pp. 103–124). Springer.
Miller, B. (2021). Is technology value-neutral? *Science, Technology & Human Values, 46*(1), 53–80.
Morrow, D. R. (2013). When technologies makes good people do bad things: Another argument against the value-neutrality of technologies. *Science and Engineering Ethics, 20*(2), 329–343.
Pitt, J. C. (2014). "Guns don't kill, people kill"; Values in and/or around technologies. In P. Kroes & P. P. Verbeek (Eds.), *The moral status of technical artefacts* (pp. 89–101). Springer.
Tollon, F. (2022). Artifacts and affordances: from designed properties to possibilities for action. *AI & Society, 37*(1), 239–248.
Winner, L. (1980). Do artifacts have politics? *Daedalus*, 121–136.

Chapter 2: Ethics of AI

Coeckelbergh, M. (2020). *AI ethics*. The MIT Press.
Dubber, M. D., Pasquale, F., & Das, S. (Eds.). (2020). *The Oxford handbook of ethics of AI*. Oxford University Press.
Hopster, J. K. G., Arora, C., Blunden, C., Eriksen, C., Frank, L. E., Hermann, J. S., et al. (2022). Pistols, pills, pork and ploughs: The structure of technomoral revolutions. *Inquiry*, 1–33.
Jecker, N. S., & Nakazawa, E. (2022). Bridging east-west differences in ethics guidance for AI and robotics. *AI, 3*(3), 764–777.
Müller, V. C. (2023). Ethics of artificial intelligence and robotics. In E. N. Zalta & U. Nodelman (Eds.), *The Stanford encyclopedia of philosophy* (Fall 2023 Edition). https://plato.stanford.edu/archives/fall2023/entries/ethics-ai/
Nickel, P. J. (2020). Disruptive innovation and moral uncertainty. *NanoEthics, 14*(3), 259–269.
Sætra, H. S., & Danaher, J. (2022). Technology and moral change: The transformation of truth and trust. *Ethics and Information Technology, 24*(3), 1–16.

Chapter 3: Technological Determinism

Alvarez, M. R. (1999). Modern technology and technological determinism: The empire strikes again. *Bulletin of Science, Technology & Society, 19*(5), 403–410.
Bimber, B. (1994). Three faces of technological determinism. In M. Roe Smith & L. Marx (Eds.), *Does technology drive history?* (pp. 79–100). MIT Press.
Ceruzzi, P. (2005). Moore's law and technological determinism. *Technology and Culture, 46*(3), 584–593.
Dafoe, A. (2015). On technological determinism: A typology, scope conditions, and a mechanism. *Science, Technology & Human Values, 40*(6), 1047–1076.
Smith, M. R., & Marx, L. (1994). *Does technology drive history? The dilemma of technological determinism*. MIT Press.
Swer, G. M. (2023). 'Blessed are the breadmakers…': Sociophobia, digital society and the enduring relevance of technological determinism. *South African Journal of Philosophy, 42*(4), 315–327.